INTRODUCTION TO

Ethology

Introduction to
Ethology
THE BIOLOGY OF BEHAVIOR

BY JEAN-CLAUDE RUWET

TRANSLATED BY JOYCE DIAMANTI

INTERNATIONAL UNIVERSITIES PRESS, INC.

NEW YORK

To Jacqueline

CONTENTS

CHAPTER 1

INTRODUCTION:
ETHOLOGY OR ANIMAL PSYCHOLOGY?

When asked to define my occupation in a nutshell, I never know whether to call myself an ethologist or an animal psychologist. The ideas of the general public, and even those of more knowledgeable persons, are, in fact, rather confused about what, exactly, these terms cover. Thus, the Zoology Department of the University of Liège recently created a course entitled "Ethology and Animal Psychology," and, at about the same time, the Psychology Department created a course which covers precisely the same material but is called "Animal Psychology and Ethology." These titles seem to indicate that different fields of activity are attributed to ethology and to animal psychology by both departments, and also that more importance is attached to ethology or, on the contrary, to animal psychology, depending upon whether one is a zoologist or a psychologist, since each, in turn, is cited first.

As a matter of fact, specialists in the field, those who deal with animal behavior, are loath to use the expression "animal psychology." They are very concerned about the connotations this term may evoke in the minds of others, scientists and laymen alike. Their apprehension is well-founded since psychology, and consequently animal psychology, has not enjoyed a good reputation in scientific circles. It has been charged that psychology does not offer all the guarantees of exactitude and reliability which are the hallmarks of a scientific discipline, and hence cannot be considered a bona fide science. This misconception arose from the fact that, until quite recently, psychology was indeed more philosophy than science. It was first taken up by philosophers and other writers exploring states of consciousness, and only later approached by scientists adhering to factual analysis with strict objectivity. The suspicion surrounding psychology is disappearing in German- and English-speaking countries, where recent advances and developments in psychology are more widely known; the suspicion persists, however, in French-speaking countries, where efforts of psychologists come up against prejudice and skepticism. This attitude reflects upon animal psychology as well, and early on in my career at an institute of general biology and zoology I found skepticism to be an everyday fact of life.

Today, suspicion of animal psychology is still justifiable to the extent that it is identified with the study of psychic activity, mental life, consciousness. These problems can, as yet, be analyzed only by introspection, and it is obvious that this method of inquiry is not applicable to animals. In fact, these phenomena—since they cannot be made objective and accessible to experimental analysis, lacking precise physiological tests—escape our investigation. It is just as vain to deny their existence in animals as it is to confirm it. And to try, in spite of all, to study or discuss them with the means currently available is to succumb to anachronistic speculation and anthropomorphism.

For these reasons, the past thirty years have been marked by a concern to give animal psychology an objective definition, designating it as the *science of animal behavior.* That also happens to be precisely the definition of ethology. Granted, in French-speaking countries, *éthologie* is assigned a much more restricted meaning, being specified as the purely descriptive study of what were formerly called animal habits. This interpretation has not been modified by the fertile developments of ethology in German-, Dutch-, and English-speaking countries, but remains faithful to the original meaning of the word established by the French naturalist Geoffroy Saint-Hilaire. It can be ascertained, however, that "ethology" in English is exactly synonymous with the German term *Tierpsychologie* by comparing the content of works published in those two languages.

Ethology, animal psychology, the science of animal behavior—all, then, amount to synonyms designating that discipline which studies the total complex of innate and acquired behavior with which an animal resolves the difficulties and problems posed by its physical and biological environment in order to survive, endure, and reproduce. The best way to define the spirit in which these questions should be approached and the methods by which they can be investigated is no doubt to review briefly the broad schools of thought which have succeeded one another, confronted one another, and finally complemented one another in the course of the last hundred years.

In the nineteenth century, combining observation and experimentation, animal psychology reached the level of true science. The *instinctivist* school was dominant then, and its most famous and productive representative was the French entomologist J. H. Fabre (1823–1915), the ten volumes of his *Souvenirs entomologiques* (1879–1908) being but a condensation of his monumental work. In Fabre's time, instinct was defined as an innate life plan leading an animal "inevitably" and "inexorably" toward a goal beyond its ken. Instinct accounted for all behavior, charting the course of an animal's life toward the goal of preservation of the individual and the species, in accordance with the modalities peculiar to the characteristic anatomical and physiological make-up of its species. "Instinct," wrote Fabre, "is omniscient in the unchanging paths that have been laid down for it." Innateness, preformation, fixity, invariability, and specificity were its qualities and characteristics. Fabre and his contemporaries based these convictions on two findings: first, that pure instinct is invariable, and, second, that instinct, incapable of adjusting, is thwarted when confronted with conditions to which it is not adapted.

The behavior of the cicada-killer wasp of Languedoc at egg-laying time serves as a good illustration in this connection. When ready to lay, the wasp sets out in search of a specific prey, a female *Ephippiger*—a grasshopper of the local vineyards—swollen with eggs. After sighting the prey, the wasp attacks, plunging its stinger into the grasshopper's thorax, thus causing paralysis of the thoracic ganglia which control the legs. Then the wasp looks for a suitable spot close by to dig a burrow, for the bulky grasshopper, paralyzed but still alive, must be dragged there. Having hollowed out the burrow, the wasp returns to its prey and ensures the paralysis of the legs with a few supplementary stings in the region of the thoracic ganglia. Then the wasp kneads and squeezes the cerebroid ganglia which are responsible for the operation of the buccal parts, thus temporarily paralyzing the grasshopper's mouth without causing any external injury. Only then does the wasp seize the grasshopper by the antennae, drag it to the burrow, stuff it in, and deposit an egg.

The cicada-killer wasp, unlike other hymenoptera which attack smaller prey, cannot fly with a large grasshopper to a fixed burrow, but must dig one on the very spot where luck has led to the discovery of a grasshopper.

A hunter first, the wasp becomes a digger afterward. It can concentrate on digging a burrow at the site of the capture, since the prey, being paralyzed, cannot escape in the meantime. The paralysis of the buccal parts protects the wasp from being bitten while transporting the grasshopper overland, an awkward and potentially dangerous journey. Finally, keeping a paralyzed prey alive in a state of reduced energy consumption guarantees a fresh food supply for the offspring at hatching. But whenever an entomologist intervenes and introduces an unexpected obstacle, instinct is incapable of overcoming it. When Fabre cut off the grasshopper's antennae, the wasp was incapable of transporting it by grasping it elsewhere.

Observations of this nature were what convinced Fabre of the programmed invariability of instinct and convinced the vitalists that behavior patterns were goal-directed. Fabre has been criticized for having failed to recognize individual learning capacities. It is only fair, however, to point out that these are much less significant and apparent in arthropods than in vertebrates. Some people consider Fabre to have been a mere naturalist describing—skillfully, patiently, and meticulously, to be sure—the behavior of insects. This estimate falls far short of a true understanding of his work, for in fact each observation was verified by experiments of great finesse. Moreover, it should scarcely be necessary to call attention to the fact that Fabre's series of experiments on sexual attraction in silk moths is the origin of a current important line of research aimed at the isolation and synthetic production of attracting substances to combat insects biologically.

One day in the laboratory, Fabre saw a female of the lesser silk moth emerge from a cocoon. He had seen but a few females of this species before, and he had never seen a male. Fabre placed this female in a wire cage on a window sill. Within a few hours, males had been attracted and were clustered round the cage. In order to determine the nature of the attracting agent, the entomologist devised a series of diversified experiments. He found that a female placed in an airtight cage attracted no males, but they flocked to an empty cage in which a female had recently been kept. Fabre carried out the same experiments using the greater silk moth. He established that sexual congregation results when the females emit an attractive substance, and he determined that the males' large antennae are the receptor organs of the chemical message. Today, we know how to synthesize and diffuse these substances, and thus can lure into insect traps all the males within a given area of farmland or forest before they have a chance to fertilize the females.

In light of his work as an observer and experimenter, Fabre is the undisputed precursor of the recent studies on insect behavior by P. P. Grassé and K. Von Frisch. Still, Fabre did not fully appreciate the teachings of Darwin, and he did not attempt to give an evolutionary perspective to his own work. Fabre did indeed emphasize that insect classification could be based on characteristics of habit as well as of structure, but this conclusion remains within the realm of taxonomy and does not reach into that of phylogeny. But rather than criticizing Fabre for what he underestimated or could not discern, we should, on the contrary, admire him without reservation, both for the model which he set for us as a naturalist and for the extraordinary records of behavior which he compiled. This is how Fabre—whom Darwin himself called "the inimitable observer"—described the difficulties of his method:

> Once the chemist has worked out his plan of research after mature deliberation, he mixes his reagents and lights his

Bunsen burner whenever convenient. He is master of time, place, and circumstance. He chooses his own time; he isolates himself in his laboratory retreat where nothing will distract him from his labors; he creates at will this or that condition as reflection may suggest. . . . The secrets of living nature, however, create much more difficult and precarious working conditions for the observer of life in action. Far from being able to arrange his own time, he is the slave of the season, of the day, of the hour, of the very moment. If an opportunity arises, it must be seized on the fly without hesitation, for a long time may pass before it comes again. And since opportunity usually knocks when least expected, nothing is ready to make the most of the occasion. On the spot, experimental materials must be improvised, tactics adjusted, ruses devised. . . . Furthermore, this chance comes only to the observer who seeks it. He must stalk it patiently for days and days. . . .

The end of the nineteenth century and the beginning of the twentieth were marked by yet another trend, the *mechanistic* school, a reaction against the instinctivist and vitalist concepts, which traces its roots to the Cartesian thesis of machinelike animals reacting like mere robots. This school originated in the work of J. Loeb (1859–1924) on tropisms in plants. Loeb extended his theory of involuntary oriented movements in plants to include animal movements on the basis of the following type of observations:

A locust moves in a straight line toward a light source, even so far as to burn itself. If it is blind in one eye, it is no longer able to travel in a straight line, but moves in circles around the light source. The movement as a whole comprises an orienting reaction and motion of translation. If light strikes the left and right sides of the animal unequally, a stronger muscular contraction is provoked on the side where the light is stronger. Thus, the locust becomes oriented on the axis of the light ray, and then equal stimulation of the left and right photoreceptors causes equal contractions of the left and right muscles, and the insect travels in a straight line. Hence, the oriented movement of the locust is the resultant of two nonoriented elements, the resultant of two reflexes.

Proceeding from this significant fact, Loeb drew broader conclusions (1889–1918): all oriented movements are tropisms; behavior patterns consist of movements which are involuntary and unadapted. Behavior, then, is but an aggregate of automatic reactions. This mechanistic concept gave rise to "reflexology," which has to its credit the instigation of numerous studies on the sense organs and perception of animals. In Loeb's day, however, his theory was the target of severe criticism because of its generalizations.

There are, in fact, oriented movements which are known to be initiated spontaneously. Tinbergen (1953) cites the following example. The grayling butterfly *(Eumenis semele)* evades a predator by flying into the sun; being dazzled, the grayling is unable to continue on a straight course. This then, is a matter of phototaxis. But if a female should happen by, this dazzled butterfly is nevertheless quite able to pursue her in a straight line. This movement, also dependent on visual stimuli, is not, however, caused by the simultaneous stimulation of two visual receptors; it does not result from the combination of two reflexes.

If one accepts the theory that behavior is a set of tropisms, one must assume that the same causes always automatically produce the same effects, and this is far from being the case. A male stickleback is incited by the red belly of another male only during the period when he is defending nest-building territory (Tinbergen, 1953a). Termites normally have a positive geotropism and a negative phototropism, but the directions of these reflexes are reversed at swarming time (Grassé, 1963). The sight of a caterpillar provokes a neck-bending reflex in a bird, and the presence of a caterpillar in the mouth provokes a swallowing reflex; but at nesting time, a bird captures a caterpillar without swallowing it, bringing it back to the brood. Therefore, it must be admitted that there is more to behavior than simply a sum of reflexes.

It was Jennings (1906 and 1923) who finally demonstrated that animals are capable of oriented movements which are learned by a process of trial and error. Studying chemotaxis and reactions to light in paramecia, he found that it is by chance alone, and not by reflex movements, that they manage to find the areas of a container most favorable to them and to avoid those which are unfavorable. At about the same time (1905), Pavlov laid down the principles and laws of conditioned reflex, the key to understanding the mechanisms of learning and adaptation. Meat extract, placed in a dog's mouth, automatically provokes the reflex of salivation because of continuous physiological association. But eventually, the mere sight of the laboratory worker who brings the food, or the sound of a bell which precedes the feeding, will have the same effect. The specific unconditioned stimulus, meat extract, has been associated with some nonspecific stimulus, and a temporary link has been established between the latter stimulus and the response. This link disappears when the association ceases to be useful, that is, when the conditioned stimulus loses its signal value by ceasing to be accompanied, i.e. *reinforced*, by the unconditioned stimulus.

However, Jennings—whose violent polemics with Loeb are famous—and the followers of Pavlov as well, committed the same error as Loeb, that of generalization. Jennings tried to reduce all behavior to learning by trial and error, while Pavlovians attributed it all to conditioned reflex. Consequently, around 1930, with the triumph of mechanistic concepts, of reflexology, of stimulus-response psychology, all finality had disappeared from behavior theories. Instinct dissolved in a flood of taxes and conditioned reflexes. To Fabre's "instinct is everything" came the retort of L. Verlaine, "instinct is nothing" *(Rech. Phil.* 1932–1933:48–61).

At the same time as a productive school of psychophysiology stemming from the discoveries of Pavlov was developing in Russia, a movement destined for equal fame was springing up in the United States—*behaviorism*. This movement arose from the desire of certain American psychologists to lay aside speculative theory and take up the task of providing psychology with an accurate methodology. As opposed to psychology concerned with states of consciousness, this school aims at objectivity. It is essentially mechanistic in that it is based on the study of stimulus-response. Its affinity to the Pavlovian is evident, but, contrary to psychophysiologists, behaviorists deliberately pass over the study of intermediate physiological factors in order to concentrate primarily on the overt manifestations of response.

The behaviorists assign paramount importance to acquired behavior patterns. In fact, from Thorndike (1898) to Skinner (1938, 1963), the tendency to interpret all behavior solely on the basis of learning mechanisms has gathered strength. All behavior patterns are seen as controlled—and on occasion modified—by their consequences. For example, Thorndike placed a cat in a cage which could be opened from the inside by activating a latch or cord. In order to induce the animal to attempt to get out, food was put in sight outside the cage. The cat performed a number of haphazard movements and by chance opened the cage. Step by step, as the experiment was repeated, the cat learned to concentrate its activity in the vicinity of the latch or cord. It gradually eliminated useless gestures, and it finally selected the single effective movement which brought the reward of egress (laws of practice and effect).

Skinner set forth the principles and laws of instrumental or *operant conditioning* (or Skinnerian, or type II conditioning, as opposed to responding, classical, Pavlovian, or type I conditioning). In a typical example of his famous mechanized tests, an animal is placed in an experimental cage equipped with a lever. Operation of the lever is followed by distribution of food. The animal discovers this relationship by chance. Pressing on the lever constitutes the *response,* and the distribution of food is the *reinforcement,* which has the effect of increasing the probability that the response will be repeated. An experiment of this kind proves that an animal is not merely subject to the influence of its surroundings, but is also capable of acting on these surroundings (hence the term "operant conditioning"); but for the animal pressing the lever spontaneously, no unconditioned stimulus will trigger this action. Since initially, before conditioning, there is no cause and effect relationship between the response and the reinforcement, the experimenter is free to choose the reaction he wishes to study and reinforce it selectively. By automating his equipment, Skinner has made it possible to achieve great continuity and strict quantitative control in experiments of long duration (cf. Richelle, 1966).

The public still often thinks that behaviorism is identical to animal psychology. As a matter of fact, behaviorism is a school of psychology which is interested in animals only as experimental material. Animals are not studied per se but only as one link in a chain of research on conditioning and learning mechanisms. Consequently, behaviorism has dealt with the limited category of traditional laboratory animals (rats, mice, cats, dogs, pigeons, rhesus monkeys). It has placed them in experimental conditions which have undergone very little change (multiple choice, puzzle boxes, memory devices, mazes, Skinner boxes, etc.). It has standardized and simplified their environment to the extreme or, on the contrary, reconstructed it in every detail. Nevertheless, animal psychology cannot ignore the theoretical and methodological teachings of behaviorism. Its principle of operant conditioning is vital to an understanding of the mechanisms of learning and adaptation; its experimental equipment, already successfully employed in other fields such as pharmacology, holds great promise for the quantitative study of certain animal behavior patterns and rhythmic activities (cf. Richelle, 1966).

This opposition between instinctivists and anti-instinctivists, between vitalists and mechanists, is the historical background against which animal psychology was to rediscover its true purpose and find its modern orientation. Animal psychology centered exclusively on the laboratory—as practiced by psychophysiologists, by biophysicists of the mechanistic school, and by behaviorists—could not satisfy zoologists, so conscious of the diversity and complexity of animal behavior. Reacting against laboratory utilization of animal material which was both limited in variety and suspect due to captivity or domestication, against extreme schematization and simplification of the experimental environment and fragmentary behavior patterns, against elaboration of broad theories from a few individual and isolated facts, trained zoologists, the biologists working in zoos or naturalists in the field, realized the necessity of returning to the naturalist traditions exemplified earlier by Fabre's work on insects. The most urgent and essential tasks claiming their attention were of a descriptive nature and consisted of undertaking the vast inventory of behavior repertoires which occur in the animal kingdom, of observing and recording them in their logical sequence—in short, plotting out ethograms.

In order to illustrate how essential it is to study behavior patterns in their logical sequence, the work of Baerends (1941) on the digger wasp, *Ammophila campestris,* is often cited. At egg-laying time, the female digs a hole in which she deposits an egg and then buries a caterpillar to serve as fresh food for the larva. She repeats the same operation for a second and even a third egg. Meanwhile, however, the first and second eggs have hatched, and the larvae's food supply must be replenished. Accordingly, the female must divide her activities among several larvae at different stages of development and with varying food requirements.

Baerends noted that the female visited all her nests in the morning, and the state of the caterpillar supply at that time determined her behavior for the rest of the day. If before the morning visit a modification was made in the number of caterpillars on hand in the holes, the female could be induced to bring either more or less food during the day; but a modification made after this morning visit had no effect on her behavior. Thus, Baerends demonstrated exactly when a digger wasp is sensitive to a certain stimulus and how long the wasp remembers it.

It is doubtful that such precision could have been achieved with the classical method of memory study by means of deferred action. This consists of permitting an animal to react only after a certain time has elapsed from the emission of a stimulus; the maximum delay tolerated between stimulus and response serves as a measure of memory. With respect to the traditional method, questions also arise as to what extent an isolated event occurring under experimental laboratory conditions corresponds to the same event occurring in its natural context, and to what extent the conclusions drawn from this isolated event in a schematic and simplified laboratory situation are applicable to the event observed in its logical sequence.

The naturalists who have adopted these new methods are generally grouped under the label of "objectivist ethologists" or "neoinstinctivists." This school is essentially European, and its guiding light is clearly the Austrian zoologist Konrad Lorenz. He is the authority who gave this school its scientific cachet. His prestige is so great, his productions so fruitful, his practical and theoretical work on behavior so influential, that his predecessors as well as his successors are brought together under his name—it comes naturally to speak of Lorenzian ethology or Lorenzian behavioral analysis.

The epoch-making period of this school centers around the years 1935–1950. But all the great discoveries made at that time were already implicit, and sometimes even explicit, in the works of three pioneers. Julian S. Huxley, a true blue-blood of the zoological aristocracy of the Royal Society, devoted himself from 1910 to 1925 to the study of courtship display in grebes, loons, and herons. The American, Charles O. Whitman, and the German, Oskar Heinroth, with all the inherent enthusiasm of amateurs, turned their energies respectively to the study of pigeons (1898–1919) and the study of Anatidae—ducks and geese (1911–1914).

At first, naturalists of this sort, ethologists—equipped with simple binoculars, sometimes with a still camera, seldom with a motion picture camera, rarely with a sound recorder—were not taken very seriously in academic circles where complicated and costly equipment was revered. But these naturalists soon assumed an enviable position on the theoretical plane, and we are indebted to them for most of the modern concepts of ethology and animal psychology. Beyond any doubt, it is these ethologists who have given new life to animal psychology, brought it back to its true purpose, and restored its rationality. In proving the complementary interrelationship of instinct, taxis, and learning, they have demonstrated the vanity of the disputes in which other schools became embroiled. By building bridges, by opening up new approaches to other sciences—biology and psychology—they have pushed the scope of ethology far beyond its original descriptive confines.

Lorenz's genius could find expression because, first and foremost, he succeeded in forming that difficult synthesis of naturalist and professional zoologist. Inured to the strictness of scientific reasoning while earning his academic degrees, enriched by a broad education in philosophy, medicine, and the natural sciences, he is also endowed with penetrating insight which was tempered by the priceless experience of passing his childhood and adolescent years surrounded by animals.

This type of scholar has not been fostered where more traditional scientific attitudes prevail—where the white-smocked laboratory technician scorns the booted naturalist; where field work is thought of as a charming but rather antiquated hobby; where the researcher parading about in shorts, binoculars dangling on his chest, his nose to the wind and his eye peeled for birds, or else crawling about in the grass hot on the trail of insects, is an object of ridicule; and where we ethologists are called emulators of Fabre—what a noble title!—but alas, with a pejorative connotation.

If we wish to participate in the rapid advance of ethology, animal psychology, behavioral science, we must assimilate the discoveries of Lorenzian ethology. Drawing attention to the work of this school, diffusing its teachings, paving the way for the understanding and acceptance of its principles—these are the goals to which this book is dedicated.

CHAPTER 2

THE ANIMAL AND THE EXTERNAL WORLD

Effector Stimuli

Any organism, whatever its nature, lives within a physical and biological environment which continually confronts it with a number of problems. An animal becomes aware of these through the intermediary of its sensory equipment. It perceives its environment from selected data filtered through its sense organs. An animal's behavior is the overt expression of its relations with the external world. It consists of the entire repertoire of innate and acquired behavior patterns with which an animal encounters and resolves the difficulties of its milieu.

Every comprehensive theory of behavior must define and measure exactly the role of external input in the selection, release, and orientation of an animal's behavior patterns. It is therefore absolutely necessary to know the field of sensory perception peculiar to each species. To disregard this would be to succumb to a new kind of anthropomorphism, tantamount to studying an animal by attributing to it a field of perception, and hence a knowledge of the environment, identical to ours, all the while ignoring its own. Sensory equipment, fields of perception, and capacities for discrimination are so diverse that it is hazardous to make comparisons and generalizations which do not take these differences and peculiarities into account. In order to determine the parameters of what Jacob von Uexküll (1928) called the *Umwelt* of an animal, that is, its personal world, the world the animal perceives, ethologists must keep informed of progress and assimilate advances in the fields of comparative sensory physiology and biophysics.

It is not sufficient, however, merely to have a knowledge of sensory capacities, that is, to know what information an animal is capable of perceiving with its particular sensory equipment. The fact of the matter is that out of all the data accessible to it, an animal can, in effect, make a selection.

By means of sensitive electrodes implanted in a cat's brain, its auditory electrical activity was recorded. Initially, a ticking metronome was placed near the cat, and the recording indicated perception of background noise and the periodic beat of the metronome. Subsequently, a mouse was introduced within the cat's perceptual field, and the cat fixed its attention on the mouse; this was expressed on the graph by the disappearance of the peaks corresponding to the beat of the metronome. The cat had in effect suppressed audition of the metronome which was, however, still within earshot. In accord with what the cat was doing, censorship of part of the input had occurred between the sense organs and the brain. Therefore, at a given moment, and depending upon an animal's activity at that moment, there is perception of but part of that which is perceptible.

A controversy which set Von Hess and Von Frisch at loggerheads more than fifty years ago is very significant in this regard (cited by Tinbergen, 1953). Von Hess (1913) had enclosed bees in a dark box equipped to show two light sources simultaneously. These lights could have either the same intensity but different wave lengths, i.e., colors, or, vice versa, different intensities but the same wave length. Although every possible combination was offered to choose from, the bees always headed for the more intense light, regardless of color. From this, Von Hess drew the correct conclusion that bees perceive light intensity and the incorrect conclusion that they do not perceive color. On the contrary, Von Frisch (1914), struck by the ability of bees to seek out blue and yellow flowers in the fields, presumed that they were quite capable of perceiving color. In order to prove it, he presented some bees with blue and yellow papers intermingled with gray papers of all possible shades from black to white, among which were surely some which corresponded exactly in light intensity to the blue and yellow papers. Without ever making a single mistake, the bees alighted selectively on the blue and yellow papers. Von Frisch concluded from this experiment that bees were indeed sensitive to color, and not to intensity.

Each scientist was both right and wrong: wrong to deny the validity of the other's conclusions and generalize his own, right within the particular conditions of his own experiment. Actually, the controversy was futile, since each researcher was dealing with quite different behavior on the part of the bees. In the first experiment, the enclosed bees are attempting to escape, and consequently they seek the more intense light source, regardless of color. The second experiment involves honey-gatherers on the lookout for flowers which can be recognized by their color. Color perception, then, is important, and in this instance specific colors, blue and yellow, are what the bees are seeking. Although beginning with the same excitatory element, light in this case, different properties of this stimulus—intensity and color—are at work in the two experiments. Thus, it is the state of an animal, its activity at a given moment, which determines the data perceived, the stimuli selected.

There is also specificity of response to stimulus. This is well illustrated by the observations of reaction chains made some time ago by von Uexküll (cited by Thinès, 1966). A mated female tick, sightless, but having a diffuse light sense in the skin, orients itself toward sunlight, climbing high up on a perch to lie in wait. It remains there as long as necessary, until it is alerted to the passage of a mammal by perceiving butyric acid emanating from the mammal's cutaneous secretions. Then, the tick lets itself drop and, with a little luck, comes in contact with its host. The impact induces the tick to move about on the surface of the mammal. When the tick finally encounters a hairless area of skin having a higher temperature, it stops there and begins to suck blood.

When the water bug *Notonecta* is waiting for prey (Baerends, 1950), the first two pairs of legs and the extremity of the abdomen detect ripples on the surface of the water caused by the approach of a possible victim. Thus oriented, the bug moves toward the source of the vibrations. Sight of the prey triggers the third pair of legs, and the bug springs upon its victim. When its jaws contact the tender parts of the prey, an attack is unleashed, and finally, when the bug perceives certain chemical stimuli, it proceeds to suck the organic juices.

Therefore, according to what an animal is doing, a selection is made from the available sensory input, and in this input there is an optimum stimulus—auditory, visual, tactile—for a given response. *Consequently, given a knowledge of an animal's behavior patterns, given a knowledge of its sensory equipment and hence its perceptible world, that is, knowing what information an animal is capable of receiving from the milieu, it is necessary to determine exactly what information plays an effective role in activating a given behavior pattern—in short, it is necessary to identify the effector stimuli.*

The Experimental Method Employing Models

When first approaching the study of effector stimuli, it may be surprising how little experimental work has been done in this field. Only a few accurate examples are available, and these deal with isolated behavior patterns in various animals. The controversy between Von Frisch and Von Hess shows how difficult experiments are on the level of behavior. An experiment concerning an isolated behavior pattern can be properly interpreted only if this behavior can be related to its logical context. For example, the color perception of bees must be considered in the light of honey-gatherers on the lookout for flowers. An experiment will be all the more valid and significant the more it respects, approaches, or considers natural conditions, as is evident from the previously cited work of Baerends on the deferred reaction of digger wasps. Successful experimentation at this level presupposes that the qualities of a naturalist—especially keen observation, insight, patience, and a warm understanding of his experimental subjects—and those of a laboratory researcher—especially critical and analytical objectivity, precision and logic in working out experimental projects—can be combined in one and the same man. Niko Tinbergen personifies this rare synthesis at its best. It is he who has done the most to develop, implement, and demonstrate the ethological method of studying effector stimuli. The experiments which he has devised and carried out, and those in which he has participated, have become true classics. Because of their innovations and the significance of the concepts they have led to, it is important to pause here and analyze them.

A most comprehensive series of experiments, and one which illustrates his method especially well, deals with releasers of sexual pursuit in the grayling butterfly, *Eu-*

menis semele (Tinbergen et al., 1942). In this species, the sexes are brought together by visual stimuli, and not by olfactory stimuli as is the case in nocturnal lepidoptera such as silk moths. Poised on the bark of a tree, a male grayling awaits a passing female. When one appears, he flies off in pursuit. If the female is not ready to mate, she continues on her way, and the male resumes his vigil. If, on the contrary, the female is ready to mate, she stops and alights, and the male proceeds to the next phase of his courtship. The question that occurred to Tinbergen is: "what are the stimuli which make the male fly off in pursuit of a female?" The question is indeed a valid one, since a male pursues not only passing females, but also butterflies of other species, various insects, birds, leaves, and even the shadows they cast on the ground!

The immediate conclusion is that chemical stimuli are not involved, and that visual stimuli are very important. In order to identify the visual stimuli which actually trigger the male's pursuit of a female or one of her substitutes, Tinbergen made an exhaustive series of paper models in which he achieved every imaginable combination of the various visual characteristics of shape, size, and color (Figure 1). In a first series of lures, the size and shape were normal, but the color ranged from black to white. In a second series, the size and color were normal, but the shape varied from a normal silhouette to diverse geometric shapes. In a third series, the shape and color were normal, but the size was variable. These models were attached to a line at the end of a fishing pole, and Tinbergen and his colleagues fanned out over the countryside in quest of male graylings lying in wait for females. The lures were presented to the males both with and without animation, and the movement varied from smooth to jerky. In sum, by presenting the models one after another, each one of the visual characteristics was placed in competition with all the others. Tinbergen was doing nothing less than applying in the field—and with what insight and inventiveness—the sacrosanct principle of experimental investigation which holds that in order to measure the effect of various factors, each of these factors must be varied in turn while the others remain constant. Tinbergen tackles problems which can be solved only in the field, and he does this with a sense of precision characteristic of a laboratory researcher. His ingenuity makes such an adaptation possible. Not less than 50,000 tests on a great number of males established that larger size, darker color, a dancing movement, and proximity are the releasers of sexual pursuit in the male grayling, shape being immaterial.

FIGURE 1. Series of models utilized in the study of visual stimuli releasing sexual pursuit in the grayling butterfly, *Eumenis semele*. Color, shape, and size are successively tested in competition with one another. (After Tinbergen, Meeuse, Boerema, and Varossieau, 1942.)

It is interesting to note in this connection that the normal brown color of the female is not intrinsically important as a releaser, for the darker the shade, black for example, the better the response. Rather, the coloration of the female is a compromise between the requirements for release of sexual pursuit and those for camouflage, imitating tree bark to elude predators. It should also be pointed out here that although male graylings proved sensitive to somber tones, they are, in different circumstances, as when searching for food, also sensitive to the blue and yellow colors of flowers. Thus, it is really the state of a male—on the lookout for a female or in search of food—which determines the type of perceptible external stimuli which will actually be effective at any given time.

This methodology, applied with such success to the study of sexual pursuit in grayling butterflies (certain additional conclusions of this study will be discussed later on), was used by Tinbergen and his colleagues and students to explore other behavior patterns as well. An equally noteworthy example is that of the pecking response of chicks of the herring gull, *Larus argentatus* (Tinbergen and Perdeck, 1950; Tinbergen, 1953b).

Newly hatched herring gull chicks beg for food by pecking at the parent's bill. In this way they induce the parent to regurgitate the food which it brings back in its crop. The parent's bill is yellow with a red spot near the tip of the lower mandible, and this spot is precisely where the chicks aim their pecks. It follows that this spot plays a role in releasing and orienting the pecking response. In order to verify this relationship, Tinbergen and Perdeck presented to some chicks several series of cardboard profiles representing an adult head and bill, incorporating various combinations of size, shape, and color. Each of the characteristics was successively put into competition with all the others (Figures 2 and 3). In the first series, the chicks were presented one model with the normal white head and yellow bill but no spot, and other similar models bearing spots of various colors. The number of responses elicited was least for the bill without a spot, and greatest for the bill with a red spot (Figure 3). In order to test the influence of contrast between the bill and the spot, the chicks were presented a series of models with gray bills bearing spots ranging from black to white. The number of responses proved to be a function of the

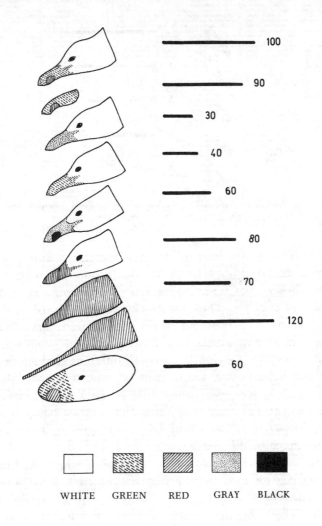

FIGURE 2. Cardboard models utilized to test the pecking response in herring gull chicks. Different characteristics of the parent's head are tested singly and then in competition with one another. The bars indicate the relative frequency of positive responses. (After Tinbergen and Perdeck, 1950; Tinbergen, 1953b).

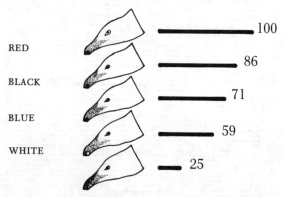

RED — 100

BLACK — 86

BLUE — 71

WHITE — 59

— 25

FIGURE 3. Number of responses to models bearing spots of different colors. (After Tinbergen and Perdeck, 1950; Tinbergen, 1953b.)

degree of contrast. In a third series, the chicks were presented decoys with bills of various colors, all without spots. Yellow was found to have no significance, but red was very effective. Then the chicks were presented models with normal bills but heads of various colors, and finally models in which the relative size of the heads and bills varied. By varying the model's different characteristics one by one—size, shape, contrast, color of head, bill, and spot—it was established that a chick does indeed react to the red spot, both color and contrast being important, whereas head and bill colors per se have no significance in this reaction.

Among altricial birds, such as perching songbirds, the young beg for food by uttering strident cries, stretching their necks and waggling their heads, opening their bills wide. Their gaping mouths display the vivid colors—yellow, red, orange—of palate, tongue, and throat, outlined by white or yellow fibrous folds. The parent automatically stuffs food down the gullet of the nearest or most demanding little beggar—first come, first served.

Fortunately, in an ample nest, a natural rotation is established. A sated chick collapses contentedly in the bottom of the nest, to digest and doze while its siblings are fed; then, when hunger gnaws once more, the chick reclaims its place, begging more and more insistently until it is fed again.

Among certain birds such as titmice, which build their nests in hollows of trees or walls, and which have broods of eight to ten young as the rule, it often happens that only two or three young, treading on their comrades, can keep a place at the entry of the narrowly confined nest to await the parent's return. If an ailing chick becomes less insistent, it is passed over by the parent and grows progressively weaker. Jostled and pushed aside by the others, its chances to claim its share dwindle rapidly, and the chick is soon doomed. The story of the swallow, "the ideal mother" fairly sharing out her pittance among her little brood, is but a moralistic legend.

The automatic manner in which the feeding response follows the begging stimulus, and the feasibility of reproducing the appropriate releasing characteristics in a model, enabled the Russians to study the diet of young nestlings (cited by Smith, 1947). Young starlings in a breeding-coop were replaced by an artificial head whose throat led down into a test tube containing alcohol. When the adult alighted on the perch at the entry of the coop, a spring was triggered which opened the model's bill and set off begging movements. This artificial system released the feeding response in the adult, and the food deposited in the model's mouth was collected in the test tube.

So far, I have cited three examples of visual releasers, but there are also other types. Numerous insects are activated by chemical means. Male moths are attracted by substances emitted by the females, and they will attempt to mate with pieces of paper impregnated with the female scent. In the grayling butterfly, after the two sexual partners have come together, it is the male which introduces olfactory stimuli during his courtship. Bowing, the male brings the scent-producing glands on the leading edge of his forewings into contact with chemoreceptors of the female's antennae. If either the scent-producing glands of the chemoreceptors are eliminated, the stage following courtship, mating, cannot take place. A hen comes to the aid of a chick upon hearing its cries of distress. If the chick is put under a bell jar, however, the hen can still see it but can no longer hear it, and she remains totally indifferent to its panicky movements. Similarly, female orthoptera—grasshoppers, locusts—are attracted by the song of distant males, but they show no interest in nearby males who are silent.

Sign Stimuli

From the preceding examples it is evident that ethologists are especially fond of birds and insects. There is also another animal which they have utilized a great deal, and which achieved celebrity along with Tinbergen. It is the stickleback, *Gasterosteus aculeatus,* a little fish very common in Holland, which is at home in both fresh and salt water. It is a very valuable research subject since its environment can quite easily be miniaturized and reproduced in an aquarium, and it therefore lends itself well to laboratory manipulations and experimental tests.

In the spring a male stickleback stakes out a territory, choosing an area of shallow water provided with aquatic plants. At this season he sports a red throat and belly, the typical prenuptial raiment. He becomes aggressive and will selectively attack neighboring territory owners cruising along his frontiers and intruders invading his domain in search of territory. These rivals are wearing the same red prenuptial raiment as the attacker, and it may be presumed that this red color enters into the release of aggression and territorial combat.

Ter Pelkwijk and Tinbergen verified this hypothesis through experiments utilizing models. As we have seen, this method consists of presenting to the animal whose reactions are being analyzed series of more or less accurate representations of the animal's counterpart in a particular activity, these models being designed both with and without the features whose releasing power is being tested. To determine the releasing effect of the stickleback's red coloration, two series of models were prepared. Those of the first series were rather vague approximations of a stickleback. Some models were very sketchy, lacking certain specific features of the stickleback and even certain characteristics of fish in general. Still other models—cylinders, spheres—were major departures from the appearance of a stickleback. All these models of series R, however, whatever their form, had their lower portions painted red. Models of series N were exact replicas of a stickleback but were neutral in color with light bellies (Figure 4). These series of models, in which the red coloration was put into competition with all the other morphological characteristics, were presented to territorial males. Either the model was introduced into occupied territory and the reactions of the defender were noted; or the defender was eliminated—thus inviting annexation of the territory by a neighbor or intruder—but replaced by a dummy, and the extent to which an invasion was prevented or delayed was evaluated. It was found that an attack was triggered or an invasion repelled exclusively by red-bellied models (series R), especially by those having oblong shapes. A model was more effective if it was also tipped downward, and in reality an aggressive male does point its snout down. So by testing the male stickle-back's reactions to characteristics such as color, shape, size, and movement, Tinbergen succeeded in identifying the releaser of territorial combat as the sign "oblong, in-clined, red-bellied object," irrespective of all the other features.

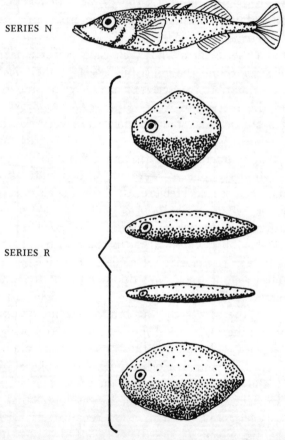

SERIES N

SERIES R

FIGURE 4. Models utilized to test releasers of fighting behavior in the stickle-back. Series N: exact models without coloration; Series R: approximate models with red belly. (After Tinbergen, 1953a.)

A female stickleback ready to spawn, her belly plump with eggs, has grey-brown protective coloration, and she keeps her head pointed upward when entering a territory. There the male greets her with a zig-zag nuptial dance, leading her in the general direction of the nest. By the same method of using models, Tinbergen showed that the releaser of the male's nuptial dance is the sign "big-bellied, oblong object" (Figure 5). The response is stronger if the dummy is tipped upward. Finally, Tinbergen showed that the releaser for spawning is a touch at the base of the female's tail, either by the male or by a small glass rod in place of the male.

FIGURE 5. Model with a swollen belly representing a female stickleback ready to spawn, releaser of the male's nuptial dance. (After Tinbergen, 1953.)

The releasing stimuli are different for each reaction. The same object may provoke two discrete reactions, but this is because different characteristics of the object are involved in each reaction. Thus, the various attributes of an object, and not the object as a whole, are what matter. The sign, and not the object bearing that sign, is what counts.

Through his use of experimental models, Tinbergen arrived at the concept of *sign stimulus* or *releaser*, expressing the idea that each animal represents a set of sign stimuli to its social and family partners, and each of these signs (for example, "oblong, red-bellied object") has the power to release a particular reaction. The animal is thus reduced to several different schematic designs related to specific responses. Much of the behavior and many of the social relationships of animals are based on this principle of specificity of response to sign stimulus. Tinbergen's masterful use of experimental models shows how this method can be employed to determine the functional value of colors, movements, structures (crests, tufts, etc.)—all the gaudy patterns of ornament and livery of a multitude of birds, fish, and insects.

Using this method, David Lack demonstrated very vividly that the red breast of the robin is the releasing stimulus for combat. A male defending its territory violently attacked a simple tuft of red breast feathers presented on the end of a wire, but remained indifferent to a neutral-colored but otherwise perfect imitation of a robin, as well as to an authentic robin whose red breast and been masked by brown dye. It is rare, however, that a single characteristic of an object constitutes such a powerful stimulus. In most of the cases studied, it is found that a response does not depend on a single stimulus, but rather on a combination of sign stimuli—to wit, the releasing combinations for the grayling butterfly's sexual pursuit, for the herring gull's pecking response, for the stickleback's territorial combat ("oblong, red-bellied object tilted downward")—and the response is more pronounced as the combination is more complete and better displayed.

At this point, we can draw two conclusions of general applicability: the first is that every reaction is closely dependent upon a very special combination of sign stimuli; the second is that each individual response in a chain of reactions is also dependent upon its particular combination, and consequently the chain comes to a halt if the stimuli do not appear in the proper sequence. Thus, a female stickleback follows a male toward the nest in response to his zig-zag dance, but she will proceed to spawn only if the male, or a substitute, touches the base of her caudal fin (Figure 6). A female grayling butterfly alights in response to pursuit by a male, but mating will not take place if the male's scent-producing glands are prevented from contacting the female's chemoreceptors. A robin defending its territory is alerted to an intrusion by hearing a rival's song, goes to meet the intruder, and, upon seeing it, attacks; broadcasting a robin's song— simulating an intrusion into occupied territory—will cause a defender to approach, but the robin will not attack the loudspeaker; an attack is released only by the appearance of a red-breasted model.

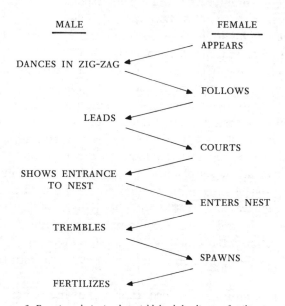

FIGURE 6. Reaction chain in the stickleback leading to fertilization of the eggs. Each response of one partner releases the appropriate response of the other partner. (After Tinbergen, 1953a.)

The Innate Releasing Mechanism

The strict correlation between a specific combination of sign stimuli and an appropriate response has led ethologists to the conclusion that a neurosensory mechanism must exist which is capable of filtering the manifold input to the sense organs, of selecting and identifying the stimulus combination present in the environment, and of releasing the correct response which corresponds to this particular stimulus situation. Lorenz called this the *innate releasing mechanism ("das angeborene auslösende Schema; angeborener Auslösermechanismus"* AAM, Lorenz, 1937b; "Innate releasing mechanism" IRM, Tinbergen, 1953a).

This is a mechanism which ensures the specificity of response to stimulus and stimulus to response. In order to explain how it operates, Lorenz used a descriptive analogy: the combination of sign stimuli is a key which unlocks a door, and one only; once opened, this door (the innate releasing mechanism) releases the corresponding response waiting within. According to this conception, each one of the responses which make up an animal's behavior depends on a particular key combination which acts selectively upon a particular innate releasing mechanism. If we pursue such analogies, we can imagine an animal's complete set of responses to various sign stimuli as being a stack of computer punch cards; the IRM peculiar to each response is a particular pattern of perforations on each card; the key combination of stimuli is the spindle action which matches the perforations and, consequently, selects the appropriate response card from the file.

This mechanism is indeed innate, for inexperienced subjects—sticklebacks and herring gull chicks raised in complete isolation, and therefore having had no opportunity to see a key combination of sign stimuli or to learn to respond—react in the appropriate manner the very first time a key combination is presented to them. After an initial encounter, an animal may very well refine and perfect its knowledge of the stimulus combination by learning—habituation or conditioning.

On the basis of the concept of innate releasing mechanisms, *releasers* have been defined as the appropriate structures, sounds, colors, odors, or movements which serve to activate the IRM. Ethologists have sometimes been criticized for defining the concept of IRM and employing the term "mechanism" without being more specific about what this mechanism consists of on the structural level, on the organic plane, from the physiological point of view. This reproach cannot go unanswered, and indeed Tinbergen (1963) has answered it very clearly. Ethologists, by virtue of their particular method of investigation, observing and experimenting on the living animal in a logical situation, have revealed the existence of a filtering system which ensures specificity of a stimulus combination emanating from the environment and an inherited motor response. When they speak of IRM, ethologists are referring to a *function*, the function of filtering stimuli and unblocking the appropriate response, without assuming anything about the structure of the mechanism or the manner in which it operates, never pretending that this mechanism should be the same in all animals and for all reactions. From another point of view, when one speaks of a leg, one thinks of the function of walking, it being understood that a leg in different animals—arthropoid or dog—may be quite different in structure and articulation. And when one speaks of an eye, one first thinks of the function of seeing, without prejudging the structure of the eye, so different, for example, in an insect and a mammal.

By demonstrating the existence of a filtering mechanism, by defining the concept of IRM, by thus posing questions as to its structure and localization, ethologists have accomplished their part of the task. It remains for others to pursue the analysis of IRM on the structural level, to localize it, to describe it more accurately and find out what makes it tick. As long as one does not interpret the concept of IRM in too strict and narrow a fashion (cf. discussions in Schleidt, 1962, and Hinde, 1966), it is an excellent tool and extremely helpful in understanding sequence and coordination in the actions of animals.

Error and Adaptation

In view of the great specificity of response and stimulus combination, in view of the close correlation which the IRM ensures between the two, and considering also that the releaser is the combination of sign stimuli and not the object or animal bearing this combination, it is quite conceivable that errors and adaptations may occur in nature, especially in interspecific relations. Error may occur on the part of a responding animal when confronted with a stimulus combination which resembles or imitates—fortuitously or not—the releasing combination of the normal social partner. Adaptation may occur on the part of the animal which bears the deceptive combination and benefits from it.

For example, a bird busy gathering food for its nestlings has been observed to approach a stretch of water and feed the gaping mouth of a fish which had surfaced to suck in air—an accidental error. The foster parents among robins, hedge sparrows, or pipits are hoodwinked when they feed young parasite cuckoos which have taken the place of their own nestlings and monopolize the parents' attention. Imitation, adaptation, and "deception" are even more pronounced among the parasitic African widow birds *(Viduinae)*. These birds lay their eggs in nests of smaller species, the waxbills *(Estrilda)*. Unlike the European cuckoo, however, a young widow bird does not eliminate its host's eggs, and consequently, the imposter grows up side by side with the legitimate offspring.

The bill, tongue, and throat of each variety of waxbill have a specific pattern of colored spots and papillae which releases and orients the parents's feeding response. A parent refuses to feed the young of a closely related variety whose gape markings do not correspond exactly to those which the parent recognizes innately. It so happens that each variety of widow bird imposes upon a particular variety of waxbill, and in begging food the young widow birds exhibit exactly the same spots, colors, cries, and movements as the young of the victimized subspecies. The error of the foster parents is flagrant, and the adaptation of the widow bird amazing! (cf. Nicolai, cited by Wickler, 1968).

By and large, in nature, the features of the head—notably the eyes—are strong stimuli evoking fear in other animals. An animal threatening another shows itself full face, but an animal making overtures to another avoids a frontal approach. Very often the frightening aspect of the eyes is softened by special structures—the eyebrows of certain birds or the diagonal stripe across the eyes of many fish which is temporarily colored by pigment cells. Many animals, however, have developed threatening features which imitate the appearance of an eye. For example, there are the opercular spots which certain cichlid fishes exhibit during combat by distending and fanning out their gill covers as far as possible. There is also the colored ocellus at the base of the dorsal fin of certain fish. The fin normally remains folded, but when these fish are exchanging threats, the fin is fanned out to display the eyelike spot. Features which have developed in imitation of eyes have also been successfully used by certain moths to repel predators. During the day, the moths rest on bark with their grey-brown forewings folded over the back. Their protective coloration and total immobility are a perfect example of mimicry. If, in spite of all, one of these moths is discovered and a bird pecks at it, the moth abandons its camouflage by abruptly flicking open its forewings to expose the hindwings, flaunting their vivid colors and big eyelike spots. Blest (1957) was able to prove the effectiveness of this sudden display of ocelli. He found that insectivorous birds, raised in cages and hence inexperienced, recoiled when a moth spread its wings—nevertheless, they did go on to eat the subjects whose ocelli had been disguised.

Wickler (1963, 1968) has studied an aggressive form of mimicry. The lipfish wrasse *(Labroides dimidiatus)* of the Pacific feeds on parasites obtained by cleaning fish of other species. When searching for food, the wrasse, already easily recognizable by its color—longitudinal black, blue, and white stripes running the length of its filiform body—signals by swimming in a jerky, bobbing manner. This set of characteristics, color, form, and movement identifies a cleaner fish to a client fish. The latter will come to a halt, turn on its side, spread its operculae, and thus invite the wrasse to proceed with the cleaning. The cleaner fish slides all along the flanks, wriggles into the gills, probes about in the mouth, everywhere rooting out skin parasites. Frequently, a client fish will actively seek out a cleaner fish and wait its turn if the latter is busy. It so happens that the saber-toothed blenny *(Aspidontus taeniatus)*, carnivorous and voracious, belonging to a different family entirely, has characteristics of shape and color and a style of swimming almost identical to the lipfish wrasse (Figure 7). Prospective clients of the cleaner fish are fooled by the mimic. They assume the position inviting cleaning and allow the saber-toothed blenny, to come within close range. Profiting from the occasion, the imposter tears off a chunk of flesh or fin and devours it.

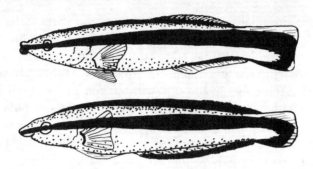

FIGURE 7. Comparison of morphological characteristics and color patterns of a cleaner fish *Labroides dimidiatus* (top) and its carnivorous mimic *Aspidontus taeniatus* (bottom). (After Wickler, 1968).

Configurational Stimuli

The use of the term "stimulus" in connection with the phenomena under discussion may oversimplify the problems of research in this area. In the case of laboratory work in physiology and reflexology, a stimulus denotes a readily definable and measurable unit of energy. It is nothing of the sort in the combinations of sign stimuli which release the behavior patterns we have studied so far. The manner in which these combinations function is very complex, as shown by the experiments of Tinbergen and Kuenen (1939) on the gaping response of young altricial nestlings.

Up to about one week of age, half of their stay in the nest, the young thrushes and blackbirds stretch their necks straight up. Not yet able to see, they are responding to a tactile stimulus, the impact of the parent alighting on the edge of the nest. After about a week, the nestlings open their eyes. They are still roused from their somnolence by the jarring of the nest, but they now crane their necks toward the head of the parent. Experiments with models have established that the characteristic which controls the orientation of the young birds' gaping is the relative size and position of the parent's head with respect to its body—regardless of shape. Any protuberance in the outline of a cardboard disk representing the parent's body attracts the nestlings' craning necks toward the bulge in the silhouette. Upon presentation of a model having two "heads," one smaller and one larger, the young orient their gaping toward the "head" whose *relative size* with respect to the disk representing the "body" corresponds to the proper size ratio between an adult bird's head and body (Figure 8). The absolute size of the head matters little; what counts is its size in relation to the rest of the body. Thus, a spatial relationship, rather than shape, is recognized, and the stimulus is *configurational* in nature.

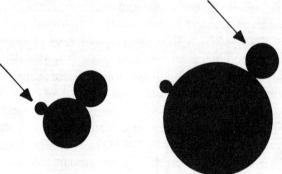

FIGURE 8. Models having two heads utilized to release gaping response in young altricial birds. (After Tinbergen and Kuenen, 1939.)

In like manner, to elicit the pecking response of the herring gull, not only is it essential that the releasing factors (spot, color, contrast) be present, but it is also important that a spatial arrangement be respected, for fewer responses are obtained if the spot—the most important structural element—is located on the head rather than on the bill. There exist auditory configurational stimuli as well. The study of reaction-eliciting elements in bird song shows that silences, intervals, and rhythms are as important as the sounds themselves.

Lorenz and Tinbergen (1938) carried out an experiment which has been repeatedly cited as a true classic. It deals with the flight reaction of birds at the approach of a predator. Ducks, geese, and gallinaceans react with alarm or flight when they see a silhouette of a bird of prey outlined against the sky. Lorenz and Tinbergen put some ducks and geese in an enclosure with an open view only to the sky. A number of cardboard models were then pulled along a wire stretched over the pen. These models were animated with a variety of movements and speeds; they represented various geometric shapes and bird silhouettes of both predators and nonpredators; they embodied various size relationships between the wings, neck, and tail of a bird in flight.

In this way the researchers demonstrated that flight response is provoked by a silhouette combining the cues "outspread wings, short neck, and long tail." This silhouette corresponds exactly to that of a bird of prey. As for the model with a "long neck and short tail," which corresponds to the silhouette of a duck or goose, it provoked neither alarm nor flight. [1] One and the same schematic model, depending upon the direction in which it is moved, can represent a bird of prey and provoke flight, or it can represent a duck and leave the test subjects to-

---

[1] Certain conclusions drawn from this experiment have been criticized, but these conclusions have no bearing on the configurational nature of the stimulus combination. Lorenz and Tinbergen concluded from this experiment that the sign stimuli (shape, speed, direction) acted upon an innate releasing mechanism. However, one can infer an innate knowledge of stimuli and automaticity of response only if an experiment is carried out on subjects raised in isolation, and this was not the case. Subsequently, tests were made on inexperienced domestic fowl, and the results showed no difference in reaction to the two types of models. These experiments are without validity, however, because they involved subjects suspected of having lost their natural flight response to birds of prey through generations of domestication. It has also been pointed out that, at

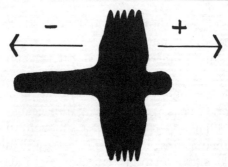

FIGURE 9. Silhouette utilized to study the alarm response of ducks and geese. Moving toward the left, the model resembles a goose. Moving toward the right, the model resembles a bird of prey and provokes the flight response. (After Tinbergen, 1953.)

tally indifferent (Figure 9). The conclusion is reached that the shape of the model is not important as such, but it is very important when related to the direction of the motion. This direction is what identifies the model as predator or duck, as dangerous or harmless. Taken together, the examples cited show that a releasing situation is contingent not only upon structural elements, but also upon the arrangement of these elements in time, space, and motion.

Heterogeneous Summation and Supernormal Stimuli

The manner in which quite different stimuli—color, size, shape, movement—combine and unite to unblock the IRM must also be examined. We have seen that sexual

first, inexperienced subjects are alarmed by anything that is new to them. Later, they gradually become accustomed to something occurring frequently and reserve their flight response only for rare, and hence unfamiliar, stimuli. Birds in the wild would retain their alarm response to predators because they appear infrequently. Nevertheless, recent experiments on young mallard ducks show significantly that naive subjects react with alarm and flight selectively to the "short neck, long tail" model corresponding to birds of prey. These experiments, then, confirm the original conclusions of Lorenz and Tinbergen, that the young birds have an innate knowledge of this configuration and possess an IRM enabling them to react appropriately at first encounter.

pursuit in the grayling butterfly, *Eumenis semele,* is induced by the following characteristics: dark color, large size, dancing flight, proximity. Tinbergen established that the strength of the response—speed and duration of pursuit—depends on the "femaleness" of the lure, that is, on the more or less complete and correct combination of various characteristics. A male grayling ultimately distinguishes a female from other objects—leaves, birds, insects—by her relative degree of femaleness. His response depends on an adequate amount of stimulation obtained from the combination of various stimuli. A deficit of one kind of stimulation (model too light-colored, for example) can be compensated for by a surplus of another kind of stimulation (larger or closer model). So the various stimuli contribute in a quantitative way to a pool of stimulation which elicits the male's response.

Studying aggression in a little cichlid fish, Seitz (1940) established that combat can be released by five different cues relating to color, size, movement, and orientation. Each of these cues presented separately on a model is almost equally effective. But the simultaneous presentation of two cues on a single model doubles the intensity of the response. In fish, as well as in butterflies, the different characteristics which make up a stimulus situation act in an additive way. Although the stimuli differ qualitatively, they have an identical effect on the response. These qualitatively different stimuli can substitute for each other quantitatively. It must be concluded that the relevant input is added up and accumulated in the nervous system, and that this combined input then acts in toto to unblock the IRM and release an adequate response. This principle is called the *law of addition* or the *law of heterogeneous summation (Reizsummenregel).*

Baerends (1962) verified this law while trying to determine which characteristics of a herring gull's egg cause the parent to recognize it as an egg and hence care for it. In this experiment, wooden eggs of various sizes, shapes, and colors (Figure 10) were placed two at a time on the edge of a nest. The female gull would first recover and return to the nest the model whose releasing characteristics were stronger. This method demonstrated the relevance of spots per se, of contrast between spots and background color, of rounded shape, and of size. It was found that herring gulls prefer speckled eggs to plain eggs, and they always choose the larger of two different-sized eggs. If a succession of choices is offered between a normally speckled egg and a plain but increasingly larger egg, the preference is at first for the speckled egg, but this is reversed at a critical size. A uniform response can be maintained by successively presenting eggs in which a decrease of one characteristic is compensated for by an increase of another characteristic—for example, an egg decreasing in contrast but increasing in size.

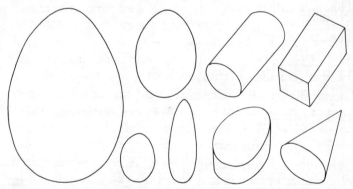

FIGURE 10. Models utilized to determine the characteristics by which a herring gull recognizes its eggs. (After Wickler, 1968.)

The way in which different stimuli can substitute for each other, and in which their effect adds up quantitatively, makes it possible to fashion models having greater releasing power than normal eggs. Such stimuli whose releasing power has been exaggerated are called *super-normal stimuli*. An example is a double-sized egg exhibiting black spots on a white background. The herring gull decidedly prefers such an egg to its normal egg. But in nature, a supernormal egg, precisely because of its strong contrast, would have but a slim chance of ever hatching, for it would be sure to attract the attention of predators. The normal herring gull egg is, then, in matter of color, contrast, and prominence of spots, a compromise between exigencies of protective coloration on the one hand and releasing requirements on the other.

Supernormality also explains the success of artificial breeding coops, which are readily accepted by certain birds, even in woods and meadows abounding in natural hollows. The fact is that each coop, nestled against a tree trunk, its size and shape sharply defined, its entry hole round and distinct, constitutes a supernormal stimulus, more attractive than a natural site to a bird in search of a hollow for its nest.

Releasing Stimuli and Orienting Stimuli

A single characteristic in a stimulus situation can be both releasing and orienting in effect. Take the case of the red spot on a herring gull's bill. This spot is the principal releasing element and it also orients the response. In many instances, however, the releasing and orienting elements are discrete. Butterflies can be induced to alight on pieces of colored paper by spraying appropriate scents in the air. The scent is the releasing element, the colored paper the orienting element. When the concentration of carbon dioxide is increased in a container of water, *Daphnia* move nearer the surface to the zone richer in oxygen, but if the water is illuminated from underneath, the water fleas move to the bottom of the container. The increase of carbon dioxide is the releasing element, and light the directing element.

The discreteness of releasing and orienting stimuli was also confirmed by studies carried out in our aquariums of the courting behavior of a little cichlid fish, the *Pelmato-chromis subocellatus* (N. Montfort-Braham and Ruwet, 1967). In this species, the female performs the principal role in the courtship dance and displays very gaudy nuptial raiment. The overall hue—head, sides, tail, fins—is yellowish; a broad black collar runs round the neck, a similar vertical band rings the tail, and a large purplish-red patch marks the underside. When a female approaches a male, he responds to her courtship by advancing and brushing the red patch with his snout. This contact is an indispensable link in the chain of movements, an essential prelude to spawning. A series of models bearing all possible combinations of the female's color characteristics (Figure 11) was systematically presented to male cichlids. It was found that only models having the two dark vertical bands elicit the male's response. But without a red patch, this response is disoriented, and the male brushes indiscriminately against the head, the side,

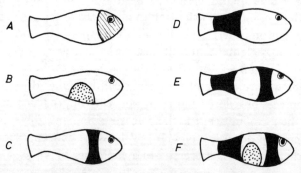

FIGURE 11. Models utilized to isolate the characteristics which release and orient the sexual response in the cichlid fish, *Pelmatochromis subocellatus.* (After Montfort and Ruwet, 1967.)

or the tail of the female model. Hence, the design of "dark collar and tail" is the releasing stimulus, and the "red patch" the orienting stimulus.

Duality of Response

Since stimulation has both releasing and orienting elements, it is not surprising that response may also have a dual nature. This was verified by the experiment of Lorenz and Tinbergen (1938) involving removal of an egg from the nest of a graylag goose *(Anser anser).* If an egg is placed outside the nest during the period of incubation, the goose stretches out its neck, wedges the egg under its bill, and rolls it back to the nest, continually making corrections when the egg tends to wobble off course due to the unevenness of the ground. If the egg is removed after the retrieving movement has begun, the movement nevertheless continues in vacuo. The retrieval of the egg, considered a single behavior pattern, breaks down into two components: (1) movement in a direct line, which, once released, must be carried through to completion even if the releasing stimulus is removed, and (2) lateral movements of correction and orientation, responding to the vagaries of the egg, which cease as soon as the egg is removed. In this case, the two components are coupled and simultaneous. In the other cases, they may occur one after the other. The bitterling *Rhodeus amarus* is a little fresh-water carp which deposits its eggs in the vent hole of a mussel. A female ready to spawn must orient herself very precisely with respect to the mussel. It is only after having achieved the correct position that, in a matter of seconds, the bitterling completes the stereotyped behavior pattern by deploying her ovipositor and depositing her eggs in the cavity of the mussel via the vent hole.

Each of these two behavior patterns—the goose's egg retrieval and the bitterling's spawning—actually consists of two elements, simultaneous or successive. One element is a *fixed component,* called a "fixed action pattern" (FAP, cf. Tinbergen, 1953), a preprogrammed series of muscular contractions which once begun must be carried through to the end, bringing the response to its *completion.* The other element is an *orienting component,* a series of reflex-type reactions to external stimuli which establish or correct the direction of the movement, a taxis guiding the response to its *objective.* This duality of instinctive response, this complementary relationship between an innate fixed element and a corrective or preliminary taxis originating from the environment, is one of the major discoveries of ethology.

CHAPTER 3

MECHANISMS OF INNATE BEHAVIOR

Spontaneity of Behavior

The previous chapter was devoted to external stimuli which release and direct behavior. I described the nature of these stimuli and how they operate, emphasizing the new concepts which ethologists have been able to formulate in this field, notably through their use of experimental models. Before ethologists defined these concepts, one school of animal psychology, the mechanistic school, had based its theory of behavior exclusively on the principle of "stimulus-response." The naturalist-ethologist school arose primarily out of a reaction against that tendency to consider animals as automatons, responding solely to the importunities and stimulation of the environment. This new school holds that the action of a sign stimuli is but one part of the mechanism which controls the expression of behavior. For an animal to manifest a behavior pattern, a specific stimulus situation does not alone suffice. An additional prerequisite is that the animal be under the influence of certain internal factors arising from its *needs*, from its *motivation*, from the *activation of its instincts*. Response requires a conjunction of motivation and releasing combination, cooperation between internal and external stimuli. Behavior is controlled both from within and from without.

A convincing example of the existence of internal influence is the fighting behavior of the stickleback. A red-bellied model provokes combat in a male only during its period of territoriality, coincident with the reproductive season. Observations of the behavior of an animal kept in constant environmental conditions, that is, subject to uniform stimulation, show that over a period of time the animal can produce a wide range of responses, from almost imperceptible reactions to reactions of maximum intensity. Stimulation capable of releasing a strong response at a given moment may have but a limited effect later on, or indeed none at all. Conversely, a standard response which at certain times requires strong stimulation may, at other times, occur under the influence of very weak stimuli—or even, in extreme cases, without any external stimulation whatsoever.

Observation and experimentation indicate that something changes with time, modifying the influence of external factors. These findings prove the existence of an *internal variable* which is also responsible for behavior. The necessary participation of internal forces constitutes a basic difference between instinctive action and pure reflex. Both of these phenomena rest on organic and genetic foundations, both respond to external stimulation perceived through the sense organs, but only instinctive action requires the additional factor of motivation. *Spontaneity, then, is an essential element of behavior.*

Pioneering naturalists like Whitman and Heinroth had already noted spontaneity. They had observed that certain behavior patterns of birds occur even in the absence of external stimuli and, hence, are due solely to the intervention of internal forces. They called such behavior patterns *endogenous movements,* since they seemed to respond to *endogenous forces.* The link between spontaneous behavior and endogenous forces is well illustrated by the following example: flies frequently clean dust particles from their wings with a fixed pattern of coordinated leg movements; wingless mutants exist which go through all the motions of grooming their wings without any possible stimulation by dust particles!

According to our current conception of the mechanisms controlling behavior, varying proportions of internal and external stimuli cooperate to elicit overt behavior. If one factor is strong, the other may be weak, and vice versa. The intensity of motivation determines the strength of stimulation required, and the strength of available stimulation dictates the necessary level of motivation.

When motivation is weak, a strong stimulus combination is needed in order to produce a standard response. There are two possibilities: either a reaction obeys the law of all-or-nothing and cannot be modified in intensity, but can be modified in the frequency of positive responses to repeated or constant stimulation; or a reaction does not obey the law of all-or-nothing and can be modified in intensity, and hence may occur in gradations ranging from a mere hint of a response to one of maximum intensity. This is why the beginning of the mating season is marked chiefly by *movements of intention*, that is, more or less sketchy responses which indicate what an animal is ready to do, what it *intends* to do.

When motivation is strong, however, a weak stimulus, or a single element of a stimulus situation, suffices to unblock the standard response. Thus, when a stickleback is defending its territory, a model having a belly barely tinged with pink provokes fighting behavior. A monkey in heat will mate with increasingly less exact female imitations. Long deprivation of food lowers the *threshold of response* to food.

The strength of internal forces determines the level of the response threshold. If the former is raised, the latter is lowered. The lowering of the response threshold due to very strong motivation can be such that the animal "explodes" and, without any external stimulus, manifests a response which is compulsive and emotional in nature. Kluyver noticed Bohemian waxwings chasing nonexistent insects during severe winters. Many birds brood before they have laid any eggs. Tinbergen (1953) observed sticklebacks perform their zig-zag courtship dance in an aquarium devoid of any females. Captive starlings, abundantly fed, nevertheless go through all the motions of predatory behavior, as if they were catching insects. And titmice, who have effortlessly gorged themselves at winter feeding stations, can be seen shortly thereafter performing all the pecking and husking actions normally necessary to obtain food—tearing up paper, cloth, etc. indiscriminately—actions with no immediate goal except that of using up the related endogenous energy.

If the motivation is maximal, then the external stimulus situation can fall to zero. Responses expressed under such conditions are called *vacuum activity* or *explosive activity.* However, since the observer is never certain of monitoring the stimulus situation in its entirety and cannot absolutely rule out the possibility of external influence, the term *overflow activity (Leerlaufreaktion)* is preferable. This implies discharge over a considerably lowered threshold, clearly the case in the behavior of the titmice cited above.

Cooperation between internal factors and external stimuli is achieved by means of an innate releasing mechanism which unblocks the reaction already primed by motivation. In the absence of appropriate external stimuli, the IRM is not activated; and if the motivation is not sufficient to cross over the response threshold alone, the reaction is not unblocked. It must be assumed that an internal mechanism accumulates endogenous impulses;

like a condenser, it stores up the internal *action-specific energy*. This hypothesis conforms to the fact that the response threshold is lowered as internal energy increases. In order to illustrate this hypothesis, Lorenz (1950) devised a "hydromechanical" or "psychohydraulic" model (Figure 12). This model does not pretend to analyze the actual mechanics of cooperation, but rather to explain

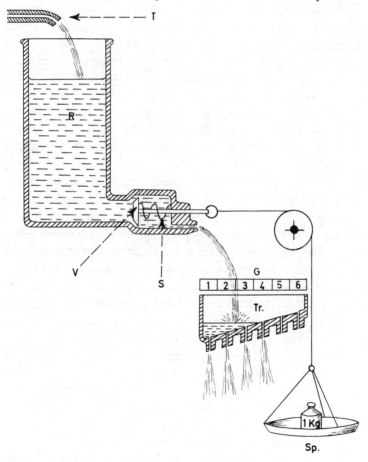

FIGURE 12. Hydro-mechanical model of Lorenz. (After Lorenz, 1950. By authorization of The Company of Biologists Limited.)

schematically, by way of analogy, the *principle of coopera-tion* between internal factors and external stimuli.

A continuous flow of water (T), representing the endo-genous source of action-specific energy, accumulates in a reservoir (R). The amount of water accumulated in the reservoir corresponds to the amount of energy on hand at a given moment for the response involved. The jet of water escaping from the reservoir represents the motor response. The force of the jet, that is, the intensity of the response, can be measured on graduated scales (G and Tr). The outlet of the reservoir is normally closed by a valve (V), which is blocked by a spring (S). The valve represents the IRM, and the spring its blocking action. The valve can be opened by the combined and variable action of the water pressure and the weight (Sp) which represents external stimuli. The water level determines the response threshold: if the water level is low, unblock-ing requires more weight, and vice versa; if the water level is very high, unblocking can occur without any out-side weight at all. Thus, the model provides a very clear and concise demonstration of cooperation between moti-vation and external stimuli. It explains graphically the principle of vacuum activity: if the internal energy is maximal, it alone can unblock the response; when the reserve of motivation has been exhausted, the motor re-sponse ceases no matter how great the external stimula-tion may be. The model also illustrates that when the valve is opened only slightly, one obtains only very sketchy responses, or else responses having a very low threshold, whereas when the valve is opened more wide-ly, one obtains responses having a higher threshold or those which register higher on the intensity scale.

The principle of cooperation brings up what Lorenz has called the "method of double quatification." This method is based upon the premise that the effectiveness of a releaser can be assessed only if the value (level) of internal factors is known, and vice versa. Testing must allow for the fact that the effect of a stimulus progressively diminishes, as measured by the intensity or frequency of response, if an animal is stimulated several times in succession with the same releaser. This decrease is not linked to muscular fatigue, physiologically speaking, but corresponds to motivational fatigue. The response wanes because each presentation of the stimulus draws upon the store of motivational energy. Therefore, in order to maintain a standard response to repeated stimulation, the intensity of this external stimulation must be increased each time.

Lorenz's explanatory model also implies that motivation can subside only through the excution of a response, since that is the only way in which accumulated motivation can be discharged from the reservoir. Manning (1967) compared this concept with the results of experiments carried out by Janowitz and Grossman (1949). These researchers operated on dogs, inserting a tube in the esophagus in such a way that either the stomach could be filled without the dog having actually eaten anything, or ingested food could be siphoned off without reaching the stomach. In this way, it was possible to dissociate the action of eating and its results. The schema of Lorenz would imply that a hungry dog whose stomach is filled artificially will nonetheless eat more food, since motivation for such behavior has accumulated and must be discharged. In reality, such a dog stops eating as soon as its stomach has been filled up, and, conversely, a dog whose stomach remains empty, the ingested food being siphoned off by the tube in the esophagus, continues to eat for a much longer time than usual. The fact of the matter is that the eating behavior of these dogs is activated or inhibited by sensory stimuli emanating from an empty contracting stomach or, on the contrary, from a full distended stomach. So it is the end result, the ultimate condition of the stomach, which matters, and, according to this concept, the psychohydraulic model, although explaining vacuum activity perfectly, would not be quite complete in that it does not provide for motivational subsidence or satisfaction other than by performance of action.

Actually, I take a dim view of any attempt to compare a comprehensive schema concerning normal animals in a logical situation with a specialized experiment on laboratory animals in a state of subjection. It is quite obvious, and also proven by the experiments of Janowitz and Grossman, that internal sensory stimuli command and countermand the ingestion of food, but the latter is only the terminal act of a complex action chain. A wild animal, feeding itself in normal circumstances, does not merely bolt down its food like a laboratory dog, but must perform a whole series of tasks preliminary to simple ingestion. It follows that an animal whose stomach has been filled without effort and for whom eating has become pointless, will be able to satisfy its motivation to eat, that is, discharge its action-specific energy by performing a posteriori all the preparatory work involved in eating behavior. So satiated animals have been known to go off hunting, and the titmice cited previously tear up paper after eating their fill at feeding stations.

Appetitive Behavior and Consummatory Action

For any given motivation, there is an appropriate response. An animal under the influence of this motivation is in a *mood*, is ready, to behave in a certain manner, and a change of motivation entails a change in mood. Usually, an animal responds only if adequate external stimuli are present. When an animal is motivated, it will either produce the appropriate response or merely give an indication of its readiness to respond, depending on whether it encounters the proper stimuli or not. In the first case, the execution of the response will have the effect of satisfying the motivation; this is spoken of as *consummatory action.* In the second case, the mood persists and will be apparent in the animal's search for the proper stimuli to unblock the response; this is called *appetitive behavior.* Thus, internally motivated behavior can manifest itself in two forms, and the terminology employed (Craig, 1918)—appetitive behavior and consummatory action—no doubt derives from analogies with eating behavior. A hungry animal has an *appetite,* being ready to respond to food-related stimuli; once the food is *consumed,* the appetite is satisfied.

When motivation is involved, three successive phases of behavior are usually discernible: an appetitive phase in which an animal manifests its mood and searches for adequate stimuli; a consummatory phase when the animal has found these stimuli; finally, a phase of subsidence or satisfaction. Motivation corresponding to a specific type of behavior—eating, drinking, mating, etc.—is frequently referred to as a *drive.*

Measuring Motivation

Investigation of an instinctive response presents the problem of finding out what role, what proportional part, is played by internal forces in the intensity of this response. Although we are able to measure external stimuli and intensity of response, we possess at present no direct means of measuring intensity of motivation. Its strength can be measured indirectly from the duration of appetitive behavior or the number of consummatory responses.

By studying an animal under constant conditions of stimulation, changes in the intensity and frequency of its
responses can be recorded. Fluctuations in the minimum
intensity of external stimulation necessary to elicit a standard response can also be measured. Moreover, variations
in the time required for a stimulus of minimum intensity
to induce this response can be measured.

Using such methods, M. M. Nice made a study of
American finches. In the spring they do not sing if the
temperature is low. But as the season advances, and as
sexual motivation and territorial drives assert themselves,
these birds will gradually tolerate lower temperatures for
singing and will frequent their favorite singing posts earlier in the morning. It can be seen, however, that this is
merely a case of indirect evaluation, for such observations
can only provide clues to the role of motivation.

A quite recent method, based on the principle of operant conditioning, seems likely to provide more accurate
measures (Richelle, 1966) and has, in fact, already produced good results. In the usual Skinner box, pressing a
lever is followed by a reward of food. This reinforcement
increases the probability that a response will be repeated.
In a modified Skinner box, pressing a lever may be reinforced, not by the usual reward of food, but by the appearance of a rival male, or perhaps a female, in an adjacent cage. Thus, the animal being studied can discharge
action-specific energy and satisfy its hunger, aggression,
or sexual motivation, as the case may be, for a length of
time fixed by the experimenter. A temporary link is established between pressing the lever and the appearance
of food or a partner, depending on the reinforcement selected. The animal will progressively come to press the
lever in order to have an opportunity to discharge its
motivation. Tabulation of standard responses and their

frequency over a period of time makes it possible to measure fluctuations in motivation.

Another experimental apparatus can be devised in which a fish passes in front of a photoelectric cell at a particular spot in an aquarium, and a relay circuit triggers a reward of food or the appearance of a male or female model. The first occurrence is entirely spontaneous, but the reinforcement increases the probability that the fish will return to the photoelectric cell and cause the model to reappear, obtaining an outlet for aggressive or sexual motivation. The experiment provides accurate measurement of aggressive or sexual drives. Godefroid (1968) perfected a technique along these lines which makes it possible to measure the daily and seasonal fluctuations of the accumulated motivational reserves of a golden hamster in varying environmental conditions.

This method involves linking a simple and measurable response to a definite drive (to eat, to fight, to mate), and then measuring the response will provide a measurement of the drive and its fluctuations. The only real difficulty is that of inventing a technique ingenious enough to link some measurable and spontaneous action of an animal to the reinforcement.

The Nature of Internal Factors

Very elementary behavior patterns are the ones most often influenced by internal sensory stimuli. The contractions of an empty stomach impel an animal to hunt; tension of the bladder wall causes an animal to urinate; a rise in the concentration of carbon dioxide in the blood, registered in the respiratory centers of the brain, induces a duck to quicken its breathing and aerate its lungs by flapping its wings, especially after diving. Hormones play an important role in controlling behavior. At reproduction time, a stickleback responds positively to appropriate models; but a castrated male, deprived of sex hormones, does not defend its territory; neither does it court nor mate. If injected with sex hormones, however, the castrated stickleback manifests normal territorial behavior.

According to Lorenz, the central nervous system produces impulses which act as a direct cause of instinctive manifestations. Neurophysiological studies of the movements of locomotion, behavior patterns on a very elementary level, have confirmed the concepts formulated about behavior patterns of a higher order. Behavior and locomotion are comprised of sequences of muscular contractions which differ only in degree of complexity. Neurophysiologists have proven that, contrary to general opinion, movements of locomotion are not merely reflexes. They include a good deal of spontaneous automatism, which is rooted in the automatic rhythms of the nervous system. Independent of any external stimuli, the nervous system produces and distributes impulses to the motor system, and continues to act autonomically even if the afferent nerves are blocked. This has been verified by experiments involving the spinal cord of a cat and the ganglia of a cockroach (Roeder, 1963). This automatic firing on the part of the nervous system obviously refutes the reflexologist thesis. The central nervous system itself is responsible for the spontaneity of behavior and is far more than a simple reflex mechanism between receptor centers and motor centers.

The impulses which are automatically fired and spontaneously distributed by the nervous system are not, however, immediately translated into movements. For that to occur, it would be necessary for an afferent stimulus either to increase the firing intensity or rhythm, or else

to remove an inhibitory mechanism blocking the expression of the primed movements. Experiments have established that the second hypothesis is the correct one:
stimuli do not increase the excitatory potential of the central nervous system, but they do unblock the impulses
produced by it.

One such experiment involved sexual behavior and
locomotion in a praying mantis, activities which usually
require very specific stimuli. When the head of a mantis
was removed, the insect proceeded to display sexual behavior in all its complexity (Roeder, 1963). The infer-
ence is that sexual behavior has an endogenous origin but
is inhibited by a mechanism situated in the head, and a
specific stimulus is necessary to remove the inhibition—
or, failing that, decapitation. This arrangement is well
adapted in the case of the praying mantis, for it often
happens that a male is decapitated upon approaching a
female! When only the supraesophagal ganglion is removed, movements of locomotion are continuously produced; when the subesophagal ganglion is removed,
unchecked sexual activity is expressed freely. One con-
cludes that these two cephalic ganglia are the inhibitory centers for these two types of activity. These facts
show that behavior originates in the organism itself,
but inhibitors normally prevent its manifestation, except
when very specific external stimuli intervene and eliminate the blocking mechanism.

The advantages of this priming system, this blocking
and unblocking, are evident. Every simple motor movement or every behavior pattern of a higher order, continually primed, is always ready to occur. But inhibitors
prevent them from all occurring at the same time and in
total anarchy. A specific action occurs only in conjunction
with a given stimulus, that is, when the action is adapted
to the situation and proves to be useful to the animal.

These processes have been studied by the neurophysiologist on the one hand, and by the ethologist on the other, each in his own way, each according to his respective methods. The neurophysiologist works in the realm of isolated organs, such as the spinal cords of cats or the ganglionic chains of cockroaches isolated from the afferent nerves. The ethologist investigates the animal as a whole within its total physical, social, and ecological environment. What is significant is that neurophysiology and ethology have arrived at the same concepts of spontaneity and of the inhibiting-uninhibiting system, one discipline studying movements of locomotion, the other studying more complex movements of innate behavior. Actually, ethologists have opened up new horizons for physiologists by making it possible for them to relate isolated experiments to the behavior of the intact animal, to put their work into context, and to express their specific results in terms of general behavior. The work of Roeder on praying mantises is an excellent example of how these two disciplines can join forces and complement each other in fruitful endeavor.

We can now see more clearly how the ethological concepts of behavior mechanisms fall in with the physiological facts. Simplified, the ethologists' explanatory schema holds that, in a suitably motivated animal, a key combination of stimuli acts on an innate releasing mechanism (IRM) which unblocks and releases a reaction, already primed and adapted to the circumstances. A better understanding of how this system actually functions is gained by analogy with what neurophysiologists know of movements of locomotion—and from which they themselves are coming to understand behavior patterns of a higher order, such as the sexual behavior of mantises. According to the neurophysiological formulation, the various innate responses in an animal's arsenal are continually primed by the automatic action of the nerve centers sending out an uninterrupted stream of impulses to the motor centers. At the motor level, each primed response is ready to be activated immediately, but inhibitors block the effective transformation of these impulses into movements. Discharge of the impulses and execution of behavior patterns occur when internal sensory stimuli and hormones lower the response threshold on the one hand, and the animal encounters the appropriate external stimuli in the environment on the other. To summarize, provided there is adequate motivation, the key combination of sign stimuli acts on the IRM, which removes the block and releases the appropriate response, the action corresponding to the accumulated impulses.

Motivational Conflicts

The correlation which exists between a key combination of stimuli and a response, and which is assured by an innate releasing mechanism, should make it possible to predict an animal's instinctive reactions with accuracy. It has been noted, however, that in the presence of a given stimulus the prediction sometimes misses the mark. The expected response does not occur; instead, an apparently aberrant behavior pattern, seemingly unrelated to the circumstances, is manifested.

Pioneers of modern ethology, like Huxley, recorded such phenomena but were unable to explain them. Two battle-ready cocks, on the point of coming to blows, eventually turn away from each other, bend over, and peck about as if looking for food. Two rival avocets, meeting on the frontier of their respective territories, threaten each other, then turn aside and assume a sleeping posture, tucking the bill under a wing. Two antagonistic starlings, at the peak of tension, suddenly begin to preen their feathers. Two territorial sticklebacks, engaged in a border fight, menace each other face to face, then turn back to the center of their respective domains and go through the motions of digging a nest. In each case, just at the moment when fighting is expected to break out, the adversaries call off the contest and engage in some activity apparently quite inappropriate to a battle situation—pecking, sleeping, preening, or digging, according to the species. In attempting to explain these phenomena, it becomes evident that something is missing in the formula, "key combination acting on the IRM which releases the response."

In still other instances, a response is clearly composite—a compromise in which components of two different responses, that is, elements related to two different instinctive drives, can be recognized. These aberrant, unexpected, composite actions occur often in the social behavior of animals which are subject to frequent conflicts between opposing drives. A territorial animal at the center of its own area is very aggressive and bellicose, but as it gets further from the center and closer to the periphery, the animal's drive to defend its territory and confront a rival progressively decreases, while its drive to avoid a rival and flee progressively increases. On the common frontier of neighboring territories, a balance of power is reached between these two drives, and consequently, it is not surprising that at this juncture mixed responses may be observed.

There is a wealth of evidence that interaction occurs between different sets of instincts. Take the case of a grazing antelope. It is hungry and grass is at hand. The animal is suitably motivated and, in the presence of the required external stimuli, it eats. But if a predator, for example a lion, comes on the scene, the antelope immediately stops grazing and flees—despite the continuing presence of all the conditions necessary for it to go on eating. It must be concluded that certain instinct systems, when activated, affect actions which are already underway. When these systems are stimulated they take priority and inhibit other systems. Obviously, certain kinds of behavior are contradictory and incompatible—eating and fleeing, incubating and escaping, hunting and sleeping—and the motivational systems responsible for them are opposing and mutually inhibiting. There are also conflicts in which no one drive dominates the others.

The venerable formula of "key combination-IRM-reaction" is not equal to the task of explaining these aberrant or composite actions, of accounting for the various possibilities of interaction between different instinct systems. A schema worked out by Baerends (1941, 1960) allows for such contingencies and provides a satisfactory explanation of motivational conflict (Figure 13). According to this schema instinct systems are organized in a hierarchical manner. Systems of a higher order and

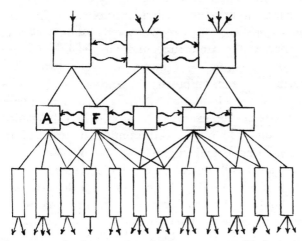

FIGURE 13. Hierarchical organization of the systems responsible for innate behavior. (After Baerends, 1960.)

on the same level with each other (A and F, for example) are opposing and mutually inhibiting (wavy arrows). A lower-ranking element may be under the influence of more than one instinct system, and a certain action pattern may result from the simultaneous activation of two different systems, responsible, for example, for actions of attack (A) and of flight (F). Due to the interplay among the different systems, the postures and movements of an animal are sometimes, or even usually, the resultant of mixed responses.

Ethologists have found this type of diagram to be very useful in explaining unexpected behavior, aberrant or composite patterns which had previously been observed but not understood. In their studies of situations in which conflicts between drives may arise, ethologists have shown that a variety of phenomena may occur in the event of interaction between two different instinct systems.

Two activated instinct systems may both be able to exert an influence. In such cases, there is said to be *ambivalence,* either simultaneous or successive. In *simultaneous ambivalence,* two motivational systems assert their influence at the same time and the result is a *compromise,* or averaging, movement, in which two behavior patterns are superimposed. According to the relative strength of the stimulation affecting each of the two activated systems—for example, approach and avoidance—a whole range of possible compromises between opposite extremes—attack and flight—may be observed. In the black-headed gull, an erect head indicates a drive to flee, whereas a bill pointing down signifies a drive to attack. When ambivalence exists between these two motivations, for example when a territorial bird encounters a neighboring rival, the black-headed gull assumes a posture of "erect head, lowered bill." As it nears its territorial border, however, the flight drive becomes stronger, its relative importance in the compromise increases, and the gull raises its bill by degrees (cf. Tinbergen, 1952a, 1959 and Moynihan, 1955) (Figure 14).

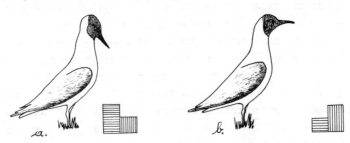

FIGURE 14. Ambivalence between attack and flight in the black-headed gull. The blocks with horizontal and vertical lines indicate the relative importance of the attack and flight drives respectively. (After Tinbergen, 1959; Moynihan, 1955; and Baerends, 1960.)

In successive ambivalence, two activated instinct systems exert their influence alternately, and the result is a *composite* movement in which components from each of the two systems occur in succession. In defending a territory during the reproductive season, a fish of the *Tilapia* family threatens neighbors and intruders by launching a frontal attack—mouth opened wide, gill covers flared out, throat puffed up, it heads for the enemy. It often happens that two neighboring fish espy each other across their common frontier and a double frontal attack ensues. As the rivals get further from the center of their territory and nearer to each other, however, the drive to attack diminishes and gives way to the impulse to flee; just at the moment the antagonists are reaching the boundary line on a collision course, they turn and retreat. Conversely, as the rivals get further from each other and nearer to the center of their territory, the impulse to flee diminishes and cedes to the drive to attack; the two adversaries head for each other again. Border skirmishes can be very lengthy affairs made up of a succession of attacking and retreating actions. Either both antagonists advance and then retreat at the same time, coming together and then separating successively, or one advances while the other retreats, and vice versa, the combined action swinging back and forth like a pendulum.

The zig-zag courtship dance performed by a male stickleback when a female enters his territory is an overt expression of ambivalence between aggressive and sexual drives. Van Iersel showed that the "zig" part of the dance is an attacking component (and it sometimes develops into a real attack); and the "zag" part of the dance is a sexual component conducting the female to the nest (and it often results in actually driving the female to the nest). By manipulating the stimulus situation through the use of models, by emphasizing excitatory factors relevant to one or the other of the two instinct systems involved, to aggression or to sexuality, it is possible to modify the relative importance of the two elements of the dance—that is, it is possible to prolong either the "zig" or the "zag."

In other conflict situations, aberrant and unexpected behavior patterns are observed which are not relevant to the instinct systems which have been activated. Kortland and Tinbergen (1940) investigated such derived activities at about the same time and called them *substitution* or *displacement* activities *(Übersprungbewegungen)*. Tinbergen has described various conditions which give rise to these activities. They occur frequently in interterritorial border fights when the fight drive is in conflict with the flight drive. The two activated instinct systems mutually inhibit each other, and the energy produced, which cannot be discharged through either of the two conflicting channels, sparks over to some other outlet offering less resistance. Thus, two avocets call off an attack and assume a sleeping posture; two rival cocks start pecking at the ground; a stickleback engages in digging, etc. It should be pointed out that the performance of a substitution activity is facilitated if the animal happens to be in surroundings offering external stimuli apt to release the substitute activity. Consequently, one can halt a cock fight by throwing grain on the ground. The rivals break off fighting and snap up the grain, but they do not actually eat it, and ultimately they just let it drop. Therefore, this substitution activity is not to be confused with true eating behavior. Similarly, substitution preening is more likely to occur if a starling's feathers are sprinkled with water.

Aside from combat situations, substitution activities may occur when sexual activity is thwarted, for example, when a partner does not appear with the expected releasers. During courtship, a stickleback burns up surplus sexual motivation by performing substitution fanning movements over the nest. Normally, the purpose of this fanning is to aerate the eggs, but in this case the movements are not made to this end since the nest does not yet contain any eggs. Still another instance of substitution occurs when a stimulus activates a drive and then ceases, and the drive can be satisfied in some other way. It is in such circumstances that certain fixed action patterns are expressed subsequent to mating. According to Tinbergen, then, a substitution activity does not result from activation of its corresponding motivational system. Rather, it proceeds from a discharge, through an unaccustomed channel, of surplus motivational energy which has been blocked elsewhere by mutually inhibiting drives or thwarted by external circumstances.

According to Van Iersel and Bol (1958), substitution activity occurs when two conflicting instinct systems are activated with equal force. These systems are mutually inhibiting and cancel each other out. At the same time, their inhibiting action on other centers decreases, and so activities corresponding to these other centers are produced. For example, incubation and flight are incompatible activities. For a bird sitting on its eggs, the very act of brooding inhibits flight. But if danger threatens, the instinctive tendency to flee increases, and in turn tends to inhibit incubation. As the danger grows imminent, a moment comes when the impulse to flee is equivalent to the impulse to brood, and the two activities are neutralized. Neither of the two can occur, but their braking action on other activities slackens, and the bird performs substitution preening. This schema of Van Iersel is a less comprehensive explanation of substitution activity than that of Tinbergen.

Motivational conflicts can give rise to still a third type of behavior. An activity may be dependent upon one of two activated instinct systems, and not completely checked by the action of the other, but merely obstructed. The activity is then reoriented, redirected toward a replacement object *(Ersatzobjekt)*. For instance, instead of striking an antagonist, a territorial herring gull may, at the last moment, reroute its movement and vent its aggression on a tuft of grass (Bastock, Morris, and Moynihan, 1953). This type of behavior is *redirection (Umrichten)*.

Comprehensive Synthesis of Innate Behavior

Tinbergen (1953a) developed a comprehensive synthesis of innate behavior based on ethological and physiological precepts. Instinct emerges at last as a reality definable in a functional sense. It is the ability to carry to completion, without prior learning, specific actions dependent on intrinsic and extrinsic factors.

External factors, or sign stimuli, can be measured, but

not without difficulty. They act through combinations—more or less configurational in nature and conforming to the law of heterogeneous summation—to release specific responses. In addition to releasing stimuli, there are orienting stimuli, which have the role of directing the response in relation to the environment.

Internal factors, the endogenous source of behavior regulating motivation qualitatively and quantitatively, are of three kinds: *hormones* and *proprioceptive sensations* increase the excitability of the sensorimotor centers and lower the response thresholds; *automatic impulses* are rhythmically produced by the central nervous system itself. These impulses are translated into movements only after being unblocked through the joint intervention of motivation and a key combination of stimuli acting on an innate releasing mechanism.

The comprehensive synthesis is based upon principles—priming, blocking, unblocking, etc.—formulated by ethologists and neurophysiologists. Identical concepts of the mechanisms of behavior were reached by the two disciplines, each working at its own level and using its own methods, ethology studying the complex behavior patterns of an intact animal in a logical situation, neurophysiology studying simple motor movements proceeding from isolated organs. There are wide variations, involving increasing complexity and more advanced integration, between movements of locomotion and a behavior pattern, but these are only differences of degree.

The most elementary level of integration is contraction of a single muscle fiber; the next level is contraction of the fibers of one muscle; then that of articulated muscles; then that of the muscles of one organ of locomotion; then the coordinated movements of the whole ensemble of organs of locomotion; and, finally, the movements which constitute what we call behavior patterns. These increasingly complex movements are dependent on mechanisms which are ever more highly integrated. These mechanisms are thought to be organized in a hierarchical system characterized by various levels of integration. Tinbergen pursues the analysis in the realm of behavior patterns, and he shows that they too reveal varying degrees of complexity and correspond to different levels of integration arranged in a hierarchical system.

The reproductive behavior of the male stickleback can serve as an example. In the spring, the reproductive mood is induced by a spontaneous cyclic change within the pituitary gland. As the days grow longer, this motivation grows stronger and the males are impelled to migrate from the sea to fresh water. They enter the estuaries, and there, in the shallow water, the higher temperature and characteristic vegetation of their breeding grounds cause the males to settle in. Each one stakes out a territory and exhibits all the manifestations connected with territory ownership—chasing off intruders, courting a female, building a nest, etc. As a group, all of these activities are dependent on the reproductive drive and are linked to territoriality, but in addition to this, each of them is individually dependent on very specific external stimuli. One cannot predict which of certain manifestations—fighting, courting, nest-building—will occur in a territorial animal unless the animal encounters the specific stimuli related to one of these activities. In the stickleback, aggression is triggered by the intrusion of a red-bellied rival, courtship by the appearance of a female whose belly is swollen with eggs, nest-building by the sight of nesting material, etc. External releasing stimuli have a limited scope, acting specifically on the mood to fight, to court, or to dig, but not on the reproductive mood as a whole.

Furthermore, although intrusion by a red-bellied rival sounds the call to battle, it does not determine which of five different types of combat will occur. The type of combat—threatening, biting, pursuing—depends upon supplementary stimuli which also have a very specific action. If the intruder bites, the defender bites in turn; if he threatens, the territorial fish also threatens; if he flees, the territory owner follows in hot pursuit. These stimuli, then, determine the type of combat, but they do not act on the fighting mood as a whole.

Different stimulus situations operate at different levels of integration. A suitable biotope induces territoriality, an intruder sets the stage for battle, and the conduct of the intruder determines the type of combat. In its reproductive behavior, the stickleback thus passes from highly integrated behavior patterns to less integrated, more fragmented behavior patterns, and these transitions are due to the action of increasingly specific stimuli (see Figure 15).

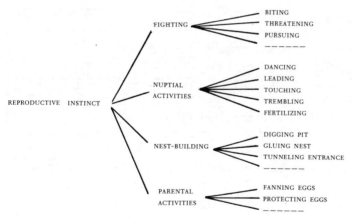

FIGURE 15. Hierarchical Organization of the Reproductive Instinct in the Stickleback

According to this concept, at the different levels of integration switches are operated by different innate releasing mechanisms complying with specific stimuli, and at each level a motor center controls a particular type of response.

Tinbergen also incorporated the concept of *appetitive behavior* into his schema of the hierarchy of instincts. Activation of a center on the lowest level of the hierarchy produces a relatively simple response, stereotyped and fixed. This response, the last of a sequence, brings about a sharp drop in motivation since it consummates the specific drives which induced the response. Activation of a center on the next higher level has two possible results: either it brings about a more marked readiness on the part of the animal to respond with consummatory acts and causes repetition of such acts in the presence of adequate stimuli; or it leads to the manifestation of random exploratory movement, not stereotyped, but flexible and variable. When a general mood like fighting or nest-building—rather than consummatory action such as biting or threatening—has been activated, an animal performs searching movements until it encounters appropriate stimuli to release the more specific consummatory act. This consummatory act is rigid in nature, being a combination of fixed action patterns and taxes. On the contrary, appetitive behavior corresponding to a general mood is variable in nature, being an "intentional" and goal-directed phase, very complex, and made up of a wide variety of fixed action patterns, taxes, conditioned reflexes, and learned behavior. To be sure, there is no real "intention" in appetitive behavior. Lorenz has stressed the fact that a goal-directed animal is not seeking a specific goal per se, but rather certain conditions in which the animal can then perform consummatory acts and thus satisfy its drives. A rat running about a maze is seeking neither its young nor food, but rather the circumstances in which it can carry out parental or eating activities.

In a similar manifestation of appetitive behavior, a hungry peregrine falcon begins hunting by circling at random over a wide area. This first exploratory phase can lead to capturing prey in various ways. The circumstances in which the predator happens upon its possible victims—a spent pigeon, a flight of teal, a scampering field mouse—determine the subsequent course of the hunt. Reconnoitering continues until the hunter comes upon a likely quarry, that is, until the falcon encounters the stimulus situation which permits it to leave off haphazard searching and pass on to a more precise phase of behavior. A flight of teal sets off mock attacks, a series of maneuvers aimed at separating a victim from the flock. This behavior is repeated until isolation of an individual duck triggers a final dive and capture. The finale consists of a chain of simple, stereotyped consummatory acts—the prey is killed, stripped, and eaten. At each level of the hierarchy, then, a center controls one type of appetitive behavior, more general on the higher levels, more specialized on the lower levels. Increasingly specific stimulus situations permit motivational impulses to pass to lower centers where they are finally consummated.

Up to the present, we have found remarkable parallels in the concepts of ethology and neurophysiology—in their ideas about the continual priming of motor movements or higher behavior patterns, and also in their principles of blocking and unblocking primed impulses. Yet another striking similarity in concept, this time with respect to the hierarchical structure of nerve centers, encouraged Tinbergen to develop his most comprehensive schema of innate behavior.

By studying the intact animal, ethologists arrived at the hypothesis of nerve centers arranged in a hierarchical order, each center performing an integrative function, receiving information and redistributing it, from the more general to the more specific. Behavior patterns, the field of the ethologist, are merely more complex and more highly integrated movements than motor movements which are the concern of the neurophysiologist. And the hierarchical organization of the nerve centers responsible for motor movements is an established physiological fact. On this basis, Weiss enumerates six levels of integration, from the lowest to the highest: (1) the single motor unit; (2) all the motor units of one muscle; (3) the coordinated functioning of several muscles related to one articulation; (4) the coordinated movements of an entire member; (5) the coordinated movements of several members; (6) the movements of the animal as a whole.

The first five levels (called Weiss 1 to 5) correspond to relatively simple movement coordinations, while level 6 corresponds to an animal's entire repertoire of behavior patterns. Tinbergen goes on, beyond the hierarchy of Weiss, to subdivide level six into several levels, each in turn more highly integrated. He provides a graphic representation of this hierarchy, and it is a remarkably comprehensive synthesis of the mechanisms of innate behavior.

Organization of an Intermediate Center

In Tinbergen's diagram (Figure 16), a center on an intermediate level (1) rhythmically produces impulses which control one kind of response. It also receives impulses from the center above (2) and is under the influence of motivational factors—hormones, proprioceptive sensations, certain external stimuli such as light and temperature—(3) which lower the response threshold. A block (4) prevents continuous discharge of impulses except when an innate releasing mechanism (5) intervenes.

The IRM is activated by a particular key combination of external stimuli (6). Impulses flow toward lower centers (7), but these centers are similarly blocked. In the absence of appropriate sign stimuli to unblock one of these lower centers, impulses are discharged temporarily by way of the appetitive behavior (8) of center 1.

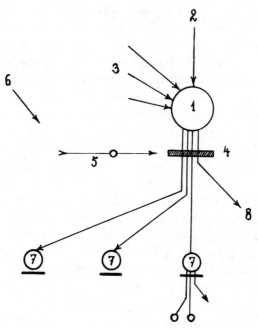

FIGURE 16. Organization of a center on an intermediate level. (After Tinbergen, 1953.)

The reproductive behavior of the stickleback can once again serve as an example (Figure 17). In the spring, the highest center, governed by the pituitary gland and stimulated by the lengthening of the day, induces reproductive behavior. Impulses produced by this center can be distributed in two ways: either they can overflow to the center on level 1, that of the major instinct controlling *reproductive activity;* or they can cause appetitive behavior specific to the highest level, that is, induce spring migration to the spawning grounds. When the fish enters fresh-water shallows, its encounter with the characteristic biotope of its species—higher temperature, specific vegetation—brings about territoriality. These key stimuli act on the innate releasing mechanism which removes the block at level 1, whereupon the level 1 appetitive behavior—oriented spring migration—ceases, and impulses discharge toward centers on level 2 controlling activity linked to *territoriality*—moods of fighting, courting, nest-building, etc. But each of these centers is blocked, so the impulses induce the appetitive behavior of level 2. The stickleback roams about its territory until a key combination of stimuli—female swollen with eggs, red-bellied intruder, or nesting material—acts on the specific IRM of one of the centers at level 2. This IRM in turn removes the block, allowing impulses to pass on to centers on level 3. If the stimulus encountered is a red-bellied intruder, the readiness to *fight* is unblocked, and the impulses induce appetitive combat behavior. This behavior is manifested until a particular action of the intruder unblocks one of several possible types of combat and causes the defender to bite, to threaten, or to pursue, to perform a consummatory act terminating the impulses.

The schema propounded by Tinbergen postulates a hierarchical organization of nerve centers controlling very precise behavior patterns. It seems that neurophysiological research is beginning to confirm this. It was already known previously that the spinal cord is the center responsible for actions situated on the lowest levels of the hierarchy, such as eating, sleeping, fighting, etc. Hess was the first to identify the hypothalamus—that region of the brain linked with the emotions—as the repository of the centers responsible for complex instinctive movements.

This research employs various types of experimental

intervention. A chemical process is used in which substances having a very specific action are injected into the brain, thus eliciting or inhibiting certain types of behavior. If a gas inhibiting aggressiveness is injected into a cat's nervous system, the cat will be terrified at the sight of a mouse.

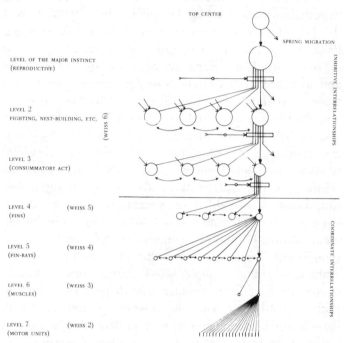

FIGURE 17. Explanatory schema of the hierarchical organization of the centers controlling innate behavior. (After Tinbergen, 1953.)

Surgical techniques can also be used to affect very precise regions of the hypothalamus. Lesions in the median area provoke continuous eating behavior leading to hyperphagia and eventually to obesity. A lesion in the lateral area, however, brings on total aphagia, and an animal will die of starvation in the midst of plenty. Thus, the centers of satiety and hunger can be assigned to dis-

crete loci in the median and lateral areas of the hypothalamus respectively. In these examples, the term "center"
denotes a nerve cell or a group of nerve cells having a
single function.

Local electrical stimulation produces effects opposite to
those of ablation. Stimulation of the median area of the
hypothalamus stops eating behavior, while stimulation of
the lateral part activates it. Von Holst and Von Saint-
Paul, using gallinaceous birds, systematized this type
of research. They made an aperture in the cranium
and attached an electrode cartridge. Several microelec-
trodes were implanted at varying depths in the diencepha-
lon. Without any external stimulation, electrical impulses
elicited behavior corresponding to the center which was
activated. In this way, Von Holst and Von Saint-Paul
produced complete behavior patterns of fighting, fleeing,
sleeping, eating, preening, and mating. Simultaneous
stimulation of two different centers produces two different responses. This may result in a compromise movement as in simultaneous ambivalence, or the two responses may occur one after the other as in successive
ambivalence. For example, one stimulus causes a bird to
stretch its neck, and another stimulus causes a bird to
ruffle its feathers. Upon simultaneous application of both
stimuli, the bird stretches its neck during stimulation and
ruffles its feathers immediately thereafter. Certain stimulus combinations elicit quite novel responses. In a hen,
separate stimulation of one center releases aggressive
pecking, and that of a second center provokes flight.
Combined stimulation produces agitation and frantic
cackling. This behavior is reminiscent of that of hens
taken by surprise near their nest, when they are torn
between two contradictory drives, the drive to flee from
danger and the drive to attack an intruder and protect
the brood. The result is a display in which the hen feigns

injury and thus draws the intruder's attention away from the nest.

These experiments confirm the hypotheses of ethologists with respect to the existence of centers responsible for instinctive behavior, the possibilities of conflict between centers, and the manifestation of mixed responses. Great care must be taken, however, in attempting to identify these centers and localize functions. Various centers affect one another. In fact, they influence one another to the point that the same stimulus applied in exactly the same place but at different times does not always elicit the same response. Despite such problems, this type of research, where the specialized experimenter induces behavior patterns which the naturalist identifies, is one of the best avenues open to fruitful collaboration between ethologists and neurophysiologists.

Consideration of their findings has led to a redefinition of instinct. It is a hierarchically organized nervous mechanism, susceptible to primary, releasing, and directing stimuli, of internal and external origin, and controlling coordinated movements which contribute to the preservation of the individual and the species. Consequently, one can no longer speak of an instinct unless these conditions are met. In this connection, there is no social instinct, for social relations are made up of a multitude of actions dependent on various centers which control attack, flight, sexuality, etc. But there is a sleeping instinct, since it has been possible to pinpoint the locus of a specific center which controls sleep and its particular appetitive behavior. *Instinct, then, emerges as a tangible reality, susceptible to localization and experimentation.*

Interaction of Internal and External Stimuli

Tinbergen's schema is extremely useful in explaining

the mechanisms of innate behavior. Comprehensive as it is, however, it still falls short of fully accounting for the machinery of behavior in all its complexity. The work of the Englishman Hinde on the reproductive cycle of the canary (in Beach, 1965) and that of the American Lehrman on the turtledove's reproduction reveal the existence of a great deal of interplay between internal and external factors. These forces act mutually upon one another, continually inhibiting, stimulating, and modulating one another.

In order to reproduce successfully, a canary must find a mate, build a nest, lay eggs, incubate them, and care for the young, in that order. These various activities must occur at the proper time and in the proper sequence. In the spring, the reproductive cycle begins internally with the reactivation of the pituitary gland which has been stimulated by the lengthening daylight. This activation of the pituitary involves the production of sex hormones in both partners. In the male, the increased level of androgen influences his sexual behavior. He sings and displays, and this courtship further steps up estrogen production in the female. External stimuli thus affect internal functions. The increased estrogen leads to development of the ovules in the female. It also initiates nest building; the female gathers dry grass and begins to fashion a nest. Due to the combined influence of internal stimuli (normal rate of hormone production) and external stimuli (accelerated rate of hormone production caused by the presence of the courting male and the sight of the nearly completed nest), the female is soon ready to mate. At this time the oviducts begin to enlarge in preparation for laying. The female also begins to shed breast feathers, exposing a bare area of skin, rich in blood vessels, which will ensure close contact between the brooding canary and her eggs. This brood patch makes the breast skin extremely sensitive to contact with dry grass, and the female, reacting to this tactile stimulus from the nest, leaves off gathering grass and proceeds to line the nest with a soft layer of feathers. Several factors—enlargement of the oviducts, elimination of the irritation caused by the grass, contact with the downy lining—bring on ovulation. Finally, contact with a certain number of eggs induces incubation.

It is hard to imagine what a multitude of experiments had to be carried out in order to unravel the respective roles of internal and external factors in developing and ordering this chain of events. In conducting the experiments, internal conditions and environmental circumstances had to be manipulated at each stage of the reproductive cycle. For example, the male is separated from the female in order to retard the secretion of estrogen. On the other hand, hormones are injected in order to induce premature development of the brood patch and untimely collection of feathers for the nest. Ovulation can be hastened by lining the nest with feathers or, on the contrary, retarded by systematically removing the feathers brought by the female. Ovulation can be prolonged and incubation delayed by removing one egg each day during the course of ovulation, or, conversely, ovulation can be halted and incubation induced by filling the nest with eggs prematurely.

By the same sort of manipulations, Lehrman reached similar conclusions in his studies of the turtledove. He demonstrated that regulation of the reproductive cycle depends upon complex and reciprocal interaction (Figure 18). When a bird is alone, hormones regulate its behavior, and hormone production is in turn affected by this behavior, and by other stimuli as well. In a situation involving a male and a female, the behavior of each bird influences the hormone levels and the behavior of the

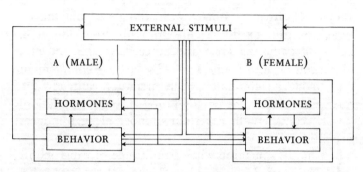

FIGURE 18. Diagram indicating interaction regulating the reproductive behavior of turtledoves. (After Lehrman, 1964.)

other, reciprocally. Stimuli emanating from A (the male) induce changes in the behavior and endocrine secretions of B (the female). These secretions in turn further modify the behavior of B. This new behavior of B causes a change in the behavior and endocrine secretions of A. This glandular change in A in turn results in a further change in the behavior of A, and so on.

Each effect, occuring on whatever level, becomes in turn a cause. Within the developing cycle, there is a great deal of interaction among the component parts, between causes and effects. These observations and experiments show that explanatory diagrams of the machinery of behavior must be elaborated to include *feedback,* the retroactive effects which play an important role in the integration, coordination, and development of behavioral sequences.

CHAPTER 4

EVOLUTION AND BEHAVIOR

Ethology and Taxonomy

Not many year ago, psychologists and zoologists tended to view questions of animal behavior from quite different perspectives. Psychologists, being primarily interested in behavior on the level of the *individual,* were particularly alert to individual variations of behavior and phenomena of learning. Zoologists, trained in a different discipline, were interested in behavior on the level of the *species.* Consequently, they focused on those aspects of behavior which are sufficiently constant and fixed to be considered characteristic of a species. They were particularly struck by the stereotyped nature of species-specific behavior, that is, innate behavior. Already in the period of Fabre and the other entomologist-naturalists, zoologists were aware of the uniformity and specificity of behavior patterns, and Fabre himself proposed identifying and classifying certain species by behavioral characteristics as well as by structural characteristics. Almost from its inception, ethology was contributing to taxonomy.

Ethology and Phylogeny

The work of Heinroth and Whitman at the beginning of this century marked a new stage in the development of ethology. Each of these ardent ornithologists chose to study his own favorite family of birds, the Anatidae (ducks and geese) and the Columbidae (pigeons and doves), respectively. They each made comparative studies of closely related species, and at about the same time, but independently of each other, they both reached certain conclusions of great significance. They showed that stereotyped behavior patterns exist which are characteristic of taxonomic groups—species, family, order, etc.—and in that respect, characteristic behavior patterns can be likened to structures or organs, can be considered in the same light and studied in a similar way. This is tantamount to saying that one can seek out and establish behavioral homologies, exactly as one does on the structural and organic levels. Ultimately, this means that one can use behavioral homologies to establish vertical and horizontal phyletic relationships. At this stage, then, ethology enters the field of phylogeny, aiding in the study of the origin and development of species.

In morphology the concept of homology dates back some one hundred years, but in ethology it is still in its infancy. Clearly, the study of behavior from an evolutionary standpoint lags far behind the study of evolutionary aspects of morphology, and for good reason. It took a very long time indeed to arrive at a theory of evolution, and longer still to chart its course using the combined resources of morphology, paleontology, and genetics. The study of the phylogenetic development of behavior patterns presents a number of special problems. To begin with, it is much more difficult to make an ethological study of an animal than to make a morphological study. The morphologist has access to collections of specimens which he may refer to whenever he wishes. The ethologist, however, must make extended observations of the living animal—intact and, if possible, in its natural habitat—before he can even begin to describe and cata- logue its complete repertoire of behavior patterns. It is not surprising that there are so few behavior inventories of species, although these ethograms are the very founda- tion of comparative research. Aside from such practical difficulties, a major problem which arises in establishing an ethogram is that of distinguishing what is learned, or acquired, from what is innate, or inherited. There is a great risk of confusing what is genotypic with what is phenotypic, a risk which scarcely exists in morphology.

Finally, the morphologist refers to paleontological findings in order to verify his hypotheses about the evolu- tion of structures, but fossil evidence provides scant infor- mation about animal behavior in times past. In morphol- ogy, comparative anatomy and paleontology are mutually reinforcing, and this makes it possible to reach solidly based conclusions. A hypothesis advanced by an anato- mist can be verified by a paleontologist. In the sphere of behavior, however, there are no such points of reference by which the ethologist can check the validity of the be- havioral homologies he proposes. So relationships estab- lished by the ethologist must agree with those established within the framework of another discipline, comparative morphology. The difficulties encountered in studying the evolutionary aspects of behavior are manifold, and this explains why comparative ethology lags relatively far behind comparative morphology.

Genetics of Behavior

After breaking new ground under Fabre, and then again under Heinroth and Whitman, the study of the evolutionary aspects of behavior reached a third stage of development with the work of Lorenz. Heinroth and Whitman held that behavior patterns could be likened to organs and studied in the same way. Lorenz, through observation and experimentation, proved that behavior patterns conform to the same laws of heredity as morphological structures. With his colleagues (Von de Wall, 1963, 1968), he studied various gestures and postures of several species of ducks, and then he crossbred the ducks in order to determine the rules by which these movements are transmitted to descendents.

In similar experiments, Dilger (1962) crossed two species of lovebirds which transport nesting material in different ways. This material consists of strips of bark, or strips of paper in the case of captive birds. One of the lovebird species carries the strips one by one in its bill, whereas the other tucks several strips at a time in the feathers of its rump. When the species are crossed, the first generation hybrids perform a sort of compromise in which elements of both methods of transport are evident. A hybrid first attempts to tuck a strip in its rump feathers, but either fails to let go of the strip or does not tuck it in properly and it falls out. The hybrid succeeds in building a nest only if it finally carries the nesting material in its bill. This inefficient two-step procedure is repeated with each strip, practice making no improvement upon the inherited compromise sequence. At the present time, the study of fruitfly mutants is being pursued by crossing varieties which differ by only a single behavioral characteristic.

Obviously, mutations affecting structures have an effect, albeit indirect, on behavior to the extent that these structures are the supporting framework of behavior patterns. In the fruit fly *Drosophila melanogaster,* the most thoroughly investigated of all animals from the genetic standpoint, it has been established that a gene called "bar" decreases the number of facets in the compound eye, and a "white" gene reduces eye pigmentation. A male fruitfly begins his courtship in response to visual stimuli emanating from a female, and impaired vision creates problems in locating females and perceiving the stimuli which induce courtship. "Forked" and "hairy" genes affect the structure and quantity of bristles. The resultant condition decreases sensibility, and those phases of courtship dependent on tactile stimuli are not well coordinated. Finally, "vestigial" and "dumpy" genes alter the shape of the wings. This results in less effective courtship and less successful mating, for the wing vibrations of a courting male are an important stimulus to the sense organs at the base of a female's antennae.

An outstanding example of behavioral genetics is the work of Rothenbuhler (1964a, 1964b) on the cell-cleaning behavior of bees. There is a certain *Bacillus larvae* which attacks bee larvae, causing death. The disease is quickly checked in some colonies where the drones open the cells and remove the dead larvae. These groups are called "hygienic." In other colonies the disease spreads rapidly, for the drones do not remove the decomposing larvae but simply abandon the contaminated cells. These groups are called "nonhygienic." Hybrids of the two groups all showed nonhygienic behavior. Hence, this characteristic is dominant, and hygienic behavior is recessive. When the hybrids were backcrossed with the recessive hygienic group, Rothenbuhler obtained astonishing results. Out of 29 colonies: (1) 6 uncapped the cells and removed the dead larvae: they were hygienic; (2) 8 did nothing: they were nonhygienic; (3) 9 uncapped the cells, but left the dead larvae in place; (4) 6 did not uncap the cells, but removed the dead larvae from cells opened by the experimenter.

Thus four types of behavior patterns were obtained in

approximately equal proportions, and it was found that hygienic cell-cleaning behavior breaks down into two components dependent on two different genes—uncapping the cell (u) and removing the dead larva (r)—which must occur together if the cleaning is to be effective. The corresponding components of nonhygienic behavior are not uncapping the cell (U) and not removing the dead larva (R). The nonhygienic alleles are dominant (UURR); the hygienic alleles are recessive (uurr). This experiment may be summarized as follows:

$$
\begin{array}{c}
1 \\
\tfrac{1}{4}\ \text{uurr} \\
2 \\
\tfrac{1}{4}\ \text{UuRr} \\
\text{UuRr} \times \text{uurr} \qquad 3 \\
\tfrac{1}{4}\ \text{uuRr} \\
4 \\
\tfrac{1}{4}\ \text{Uurr}
\end{array}
$$

Lorenz gave the initial impetus to this line of research, and the links between ethology and genetics have since become firmly established. These connections continue to be strengthened and clarified by numerous ongoing studies in this ever-expanding field. As far back as 1937, Lorenz, recognizing that stereotyped behavior patterns obey the laws of heredity, used the term *Erbkoordination*, inherited coordination. In English, these hereditary movements are called *fixed action patterns* (FAP). These manifestations of innate behavior—made up of fixed patterns of motor movements—are preprogrammed sequences of muscular contractions, genetically determined and expressed as a whole, en bloc. They are the foundation stones of comparative ethology.

Mechanisms of the Evolution of Behavior

Ethology first ventured into taxonomy, then expanded into the field of phylogeny, and most recently has found a firm footing in genetics. These are milestones in the exploration of the evolutionary aspects of behavior. Relying on its own means, ethology has independently retraced the classic route followed by morphology in the last century. Just as comparative anatomy had shown the multiplicity of structures and demonstrated their evolution, comparative ethology has revealed the diversity of behavior patterns and established their phylogenetic development.

Evolution, whether morphological or behavioral, is adaptation brought about by selection pressures of the environment. Individual members of a large population have slight genetic differences. The species makes various trials, so to speak, at the level of the individual. Natural selection acts upon these differences by favoring those individuals best adapted to the environment. It acts on, as well as through, genetic variability, so that variability is continually maintained and recreated. The environment keeps changing too, and new conditions are also encountered as a species expands its distributional range. Hence, adaptation of a species to the environment, brought about by selection pressures of the environment acting on individual variability, is a continuous phenomenon. Since the species adapts, the species changes. Its evolution reflects a close interrelationship: the environment acts upon the animal, but the animal also acts upon the environment; in changing its habitat, an animal changes the selection pressures acting on it. So *variability* and *selection* are intricately interwoven, and their consequences are *adaptation* and *evolution*.

The Adaptive Aspect of Behavior

Since natural selection is continually acting on behavior patterns, always favoring better adaptations to the environment, it may be propounded that a particular behavior pattern occurring at a given moment must be advantageous to the animal and contribute to its survival. Stating problems in terms of *function* is a characteristic peculiar to ethological inquiry. To the classic questions of the physiologist: "How does that operate? What are the *mechanisms* which release, orient, maintain, and terminate such and such a behavior pattern?" the ethologist adds the queries: "Why does the animal behave in this way? What is the *function* of this behavior pattern? How is it useful to the animal? What advantage does the animal gain from it?"

In posing questions of this nature, ethologists have come to define certain structures and behavior patterns in terms of function. Thus, a releaser is defined as an agent adapted to the function of sending out stimuli to which conspecifics respond in an appropriate manner, that is, in a manner which is advantageous to the species. This approach is considered valid, especially since Tinbergen (1963) has shown that the functions of behavior are as susceptible to experimentation as the mechanisms of behavior. Besides, when an ethologist inquires into the advantageous effect of some behavior pattern expressed at a given moment, he is, from the ethological point of view, dealing with a cause. What happens in the present is a cause which will have an effect in the future. By altering what occurs, by experimenting on the level of the cause, one alters the consequences, and the latter can be measured. This viewpoint and approach are in fact quite recent, and as yet very few experiments have been carried out in this vein. Once again, we are indebted to Tinbergen's school for the sole verifications which have been made in this line of research.

One such inquiry into the function of behavior involved certain species of twig caterpillars. When on a branch, this caterpillar assumes a very rigid pose and looks exactly like a projecting twig. It was possible to prove that this behavior is adapted, that it enables the caterpillar to escape detection by insect-eating birds. Two groups of artificial caterpillars were fashioned, one in poses to resemble twigs and the other in nonimitative positions. The results of this experiment showed that the latter group was considerably more subject to predation by birds.

Another investigation was carried out in a densely populated colony of gulls. All the females in such colonies lay their first egg on almost the same date. This simultaneity, which is assured by physiological rhythms and synchronized breeding, is adapted. It reduces the risk of the eggs being destroyed by corvine predators. This was verified by placing artificial eggs in nests at earlier and later dates. Most of these eggs were carried off or destroyed by crows. Predation, then, is a selection pressure which determines ovulation throughout the colony at about the same time. The advantage lies in the fact that all the pairs of the colony are at the same stage of the reproductive cycle at the same time, thus assuring a common defense of the eggs and chicks against intruders. If a gull were to have a different pituitary rhythm inducing earlier ovulation, this trait would have little chance of being transmitted to the next generation, for predators would destroy her eggs.

A third example shows how a behavior pattern can be a compromise resulting from two or more selection pressures. After her chicks hatch, a black-headed gull disposes of the broken eggshells by taking them in her bill and flying beyond the colony's nesting grounds to discard them. She waits about two hours after the chicks hatch, however, before cleaning the nest. That happens to be how long it takes for the newborn to dry off. It was proven that corvine predators tend to single out nests containing eggshells. It was also found that neighboring gulls are likely to attack a newly hatched chick whose down is still wet. Removal of the eggshells two hours after hatching takes both of these hazards into consideration. Immediately after hatching, intraspecific predation is the greater risk; two hours later, this risk is less than that of attracting marauding crows. The total nest-cleaning behavior—a two-hour delay, then removal of the eggshells—is the result of two selection forces acting together.

Multiple pressures are present everywhere, and the effects of their action are particularly evident in the flamboyant finery and display of courtship behavior, which has the function of attracting the attention of females or rivals, but inevitably attracts the attention of predators as well. During the reproductive season, a species derives an advantage from developing these ornaments and displays, even at the risk of attracting predators. But it is in the interest of the species to limit the duration of these showy exhibitions of movement and color. In birds characterized by vivid colors and pronounced sexual dimorphism, this limitation is achieved through pre- and postnuptial moulting. In fish, the problem is solved by means of pigment cells which make coloration patterns highly unstable. A cichlid fish can change without transition from gaudy patterns to protective coloration and vice versa, depending on whether a female appears or danger approaches.

A behavior pattern can almost always be said to be adapted. Some action which may seem aberrant in a captive animal becomes clear and meaningful when observed in natural surroundings. In a given environment, behavioral similarities among different species may be detected. These resemblances may indicate affinities among these species. On the other hand, they may result from convergence, in which case the behavior in question is undoubtedly advantageous in this particular environment. For instance, most passerine birds assert their territorial claims by singing from a high perch. But species such as larks, pipits, and fan-tailed warblers, which breed in treeless steppes, heaths, and prairies, have all developed a territorial flight accompanied by song. This flight is a behavior pattern adapted to an environment devoid of perches and serves to enhance the territorial song.

There are some examples of behavior—substitution, vacuum, or displacement activities—which appear to be completely aberrant, inappropriate. Although these expressions of behavior may seem unadapted, we know that they result from the action of behavioral mechanisms, and these are adapted. It is indeed important for an animal to release its surplus motivation, to discharge impulses whose normal egress is blocked, to relieve a state of tension. Similarly, the hierarchical organization of instincts (Tinbergen, 1953a)—by channeling impulses into appetitive behavior or toward subordinate centers —performs the important functions of ordering an animal's progress from very general behavior patterns to ever more specific ones, and of forcing the animal to delay the consummatory act until the action is perfectly adapted to the situation. The very machinery of behavior, itself adapted, assures the adaptedness of overt behavior. Consequently, in order to determine to what extent a behavior pattern is adapted, it is not only necessary to study the animal as a whole, but it is essential to comprehend the situation as a whole, its physiological as well as ecological aspects. And then the two questions, "How?"

and "Why?" fuse into one: "How does an animal behave in order to survive?"

Findings of Comparative Ethology

I have already enumerated the difficulties comparative ethology comes up against in making inventories and establishing homologies of fixed action patterns. Now I shall turn to the first results of these efforts. Only a few zoological groups have been studied from a comparative standpoint. These are: pigeons (Whitman, 1919); Anatidae—ducks and geese (Heinroth, 1911; Lorenz, 1941; Von de Wall, 1963, 1968); Laridae—various gulls (Tinbergen, 1959); spiders, mantises, and fiddler crabs (Crane, 1949; 1952; 1957); grasshoppers (Jacobs and Faber); and cichlid fish (Baerends et al., 1950, 1963). Precisely because such studies are so rare, and so valuable to the development of theory, it may be interesting to present a few examples in detail.

Laridae. Over a period of several years, Tinbergen and his colleagues studied a great many species of gulls in their natural habitats in various parts of the world. These birds constitute a homogenous morphological group, the Laridae. They also all share a common heritage of behavior patterns, by which each and every one of them can be recognized as belonging to the Laridae, just as surely as by morphological criteria.

Most of these species, which live in coastal regions, nest on the beaches. One species, however, the black-legged kittiwake, has adapted to a particular environment. The coasts and islands where this gull lives are characterized by cliffs, and it nests on narrow ledges high above the ocean. Through comparative anatomical and ethological studies, Cullen (1957) has found thirty-three ways in which the black-legged kittiwake has come to differ from other gulls, although they have a common Laridae inheritance. Laridae characteristics can be clearly recognized in the kittiwake's physical attributes and behavioral traits, just as a mammalian forelimb is recognizable in a bat's wing. Due to the kittiwake's particular way of life, however, in a habitat peculiar to this one species out of the entire group, this gull has put its personal stamp on the behavior patterns which are a part of the common heritage of all laridae.

I shall cite two examples of behavioral adaptation in the kittiwake. The majority of gulls assert their territorial claims by "long calls" and an "oblique posture," and they indicate a nesting site by "bobbing" the head and body toward the ground (Figure 19). In the kittiwake's habitat, territory and nesting site are merged into one

FIGURE 19. Postures used by gulls to defend territory and indicate a nesting site. (After Tinbergen, 1959.)

small spot on the narrow ledge of the cliff, and this gull uses but a single movement, "bobbing," to indicate the nesting site and also to defend its territory. Thus, one inherited action pattern of the Laridae, superfluous in these particular circumstances, has been lost. It is appropriate to note here that an alteration in behavioral inheritance usually involves the loss of some element, it being easier to outgrow an action pattern than to form a new one. In behavior as in morphology, once an element is lost in the course of evolution, it may be replaced, but never recovered.

Figure 20. Appeasement gesture of gulls, turning the head aside. (After Tinbergen, 1939.)

In typical gull behavior, two adult birds which come face to face in a combat or pairing situation will appease aggressiveness in their opposite number by turning the head aside (Figure 20). A threatened chick, however, runs off in alarm to hide. But a kittiwake chick, confined to its narrow ledge, cannot run for cover, so when it is endangered, it resorts to the same action pattern used by rival adults—it turns its head aside in the ritualized gesture of appeasement. This precocious expression of a behavior pattern typical of adult Laridae is how the kittiwake adapts to one of the pressures of its particular environment. Although these inherited behavior patterns can be adapted to special situations, the action patterns remain very constant and traditional in form. They keep their basic character and, even when they are used in a new context or assigned a new function, it is still possible to recognize their origin and identify them as Laridae characteristics.

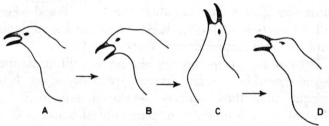

LESSER BLACK-BACKED GULL

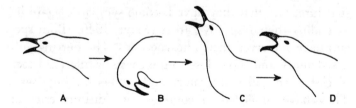

HERRING GULL

FIGURE 21. Differences of form and amplitude in the movements used by the lesser black-backed gull and the herring gull to indicate their presence in their territory. (After Tinbergen, 1959.)

Under the selective force of environmental pressures, fixed action patterns which are inherited in common by closely related species may come to differ in two respects: form, on the one hand, and frequency of use, on the other. The territorial behavior of the Laridae—"oblique long calls"—has a very constant basic form throughout the group (Figure 21): initially a gull lowers its head (a-b); then it throws its head back and calls (b-c); finally the head is returned to a normal position (c-d). Each species, however, while following the same basic sequence, stresses and highlights different parts of the sequence. The behavior pattern has undergone a sort of microevolution. Although the elements of the sequence derive from a common origin, the sequence as a whole has assumed a different style or rhythm in different species. These variations in form make it possible to distinguish one species from another: the lesser blackbacked gull tosses its head further back, while the herring gull bends its head further down (Tinbergen, 1959).

These same two species can also serve to illustrate the second type of behavioral variation, frequency of use. It is thought that these species developed separately in different areas, from different geographical branches of a single species; that they underwent divergent evolution which has left its mark on their behavior as well as their structure; and that they have become sympatric again by extending their respective areas (Mayr, 1940). Both species have preserved their idiosyncrasies. The herring gull is sedentary and stays close to the coasts, while the lesser black-backed gull is migratory and prefers the high seas. They have inherited in common two distinct cries of alarm. One of these, used more often by the herring gull, expresses alarm of low intensity, while the other, favored by the black-backed gull, expresses alarm of high intensity. Sometimes these gulls nest in mixed colonies and, in

identical surroundings, the same source of anxiety provokes the low-intensity call in the herring gull and the
high-intensity call in the black-backed gull. There is,
then, a difference between the two species in the response threshold for alarm. As yet, this is only a
difference in frequency of use, but the disappearance in
each species of the less frequent response is conceivable.
The loss of this characteristic would create a difference in
the behavior repertoires of the two species.

Anatidae. Since evolution can bring about changes in
the behavior repertoires, it can be seen that once an inventory has been completed of the innate action patterns
of a zoological group, a family for example, the presence
or absence of certain characteristics may have taxonomic
significance. Lorenz (1941) made a comparative study of
the Anatidae family (geese, barnacle geese, sheldrakes,
ducks), and then used the ethograms of the various species in taxonomy. He worked out a conspectus in which
each species is classified according to its stock of action
patterns. The study shows that certain behavior, such as
the monosyllabic peeping of a lost chick, is common to all
Anatidae; other behavior patterns are held in common by
ducks, but are nonexistent in geese; and still others are
characteristic of only certain groups of ducks. The use of
these behavioral criteria, when traditional taxonomic
criteria have been exhausted, has made it possible to
refine and adjust the classification of certain species.

In his study of the Anatidae, Lorenz (1958) also in-
ventoried, catalogued, described, and named a series of
gestures which are part of the common heritage of surface-feeding ducks, mallard, pintail, gadwall, teal, etc.
These movements are peculiar to the drakes, and are performed in the water in a group display during courtship.
This betrothal ceremony culminates with the selection of

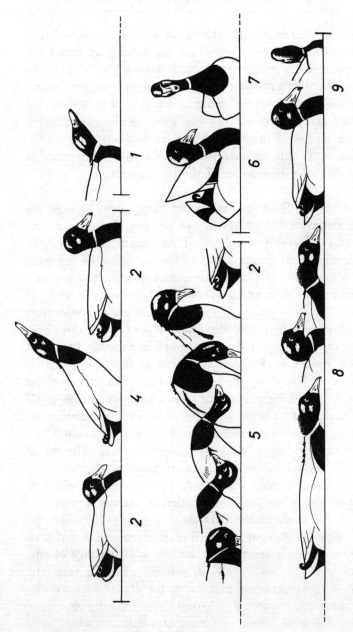

FIGURE 22. Courtship movements of surface-feeding ducks. The sequences shown here are peculiar to mallard drakes. (After Lorenz, 1958.)

one of the males by a female. These courting movements are, among others (Figure 22): (1) *bill shake:* the bill is shaken in a vertical plane; (2) *tail shake:* the tail is shaken in a horizontal plane; (3) *seesaw:* very rapidly, the drake dips forward, plunges its bill in the water, then draws its head as erect as possible, still keeping the breast low in the water and the tail tipped up; (4) *shake and stretch:* the head is drawn back within the shoulders and shaken rapidly two or three times, then the neck shoots out, raising the entire body above the water line; (5) *water spout:* the duck plunges its bill in the water, sends a spray of water off to the side, riffles the water again with its bill, then stretches full height above the water, and, finally, resumes a normal position; (6) *high and short:* the head is drawn up and back and the tail held erect, giving an impression of increased height and foreshortened length; (7) *turning forward:* the drake turns its body toward a female; (8) *swimming flush with the water:* the duck swims with its body flattened on the water, accompanied by forward thrusts of the head; (9) *turning away:* the drake turns away, showing the back of the head to the female.

These movements, which sometimes differ very slightly in form from one species to another, can be considered homologous among the various surface-feeding ducks. Each species, however, integrates these movements into its courtship display in a particular order. When a female mallard alights near a group of drakes, they perform three principal action sequences in succession. First they perform "introductory shaking" invariably made up of movements 2-4-2. Then they do a sequence made up of movements 1-5-2, in which a "water spout" is preceded by a "bill shake" and followed by a "tail shake." Finally, if the female still remains unmoved, they go through an elaborate sequence composed of movements 6-7-8-9.

Teal courtship also has three main sequences made up of the same basic elements. Once again, the group display begins with "introductory shaking" 2-4-2. The second act consists of several seesaw movements, 3, performed in rapid succession. The climax is reached with a still more elaborate sequence than that of the mallard, composed of movements 5-2-4-6-7-9 in that order. Depending on how highly the drake—teal or mallard—is motivated, the third sequence may be executed in its entirety, or it may cease at almost any point. In both species, however, certain movements (6-7) invariably follow each other, being bound together so inextricably as to form but a single action pattern.

Combining commonly inherited elements into different sequences evidently imbues the courtship of each species with a special meaning understood by that species alone. Although fertile hybrids are readily produced by surface-feeding ducks raised in the promiscuous conditions of captivity, this cannot happen in the wild. These courtship displays, initially based on elements common to the different species, have become so specific that in effect they prevent crossbreeding. They maintain separation among various species living in the same habitat, acting as an ethological isolating mechanism. The courtship sequences are very specific, but the constituent elements of these displays can still be identified beyond question as characteristics of surface-feeding ducks in general.

Cichlid Fish. Baerends and Baerends-Van Roon (1950) laid the groundwork for the comparative ethology of cichlid fish. Since then, the cichlid group has been and continues to be the subject of extensive research in a number of laboratories and aquariums. The various species show a common inheritance of action patterns by which it is possible to identify a cichlid fish with as much certainty as by the usual morphological criteria. Comparisons of behavioral makeup—a fish's repertoire of movements, the form of these movements, their frequency of use, the context in which they occur—also makes it possible to distinguish taxonomic groups such as genera, subgenera, and species. There is no doubt that comparative ethology is already making very valuable contributions toward clarifying the taxonomy and phylogeny of this rapidly evolving group. However, this study is a tremendous undertaking, for the cichlid family is immense, comprising an enormous number of species, and still new ones are constantly being reported.

Ritualization

Fixed action patterns are very stable and resistant to change, subject only to minor modifications. This constancy of form is a particularly useful research tool since it makes it possible to recognize and verify the unity of a taxonomic group. But although every action pattern is inherently resistant to change, there is, nevertheless, a certain level of behavior where action patterns yield more easily to selection pressures, where rapid evolution and diversification can occur.

Social relations among animals abound in relatively malleable action patterns, and the behavior expressed on this level might be said to be emotionally motivated, since it is related to fright, aggressiveness, sexual approach, or varying combinations of these. A particular behavior pattern may reflect a single one of these emotions, for example, fright. The typical posture of a frightened bird on the point of escaping by flight is flexed legs, lowered body, extended neck, outspread wings—ready for take-off. An aggressive bird on the point of attacking will ruffle its feathers, thrust out its head, threatening the intruder with its bill. The point pertinent to this discussion is that these actions may be quite vague and sketchy or may be very bold and pronounced, according to the intensity of fear or aggressiveness which has been aroused in the bird. There is a whole gamut of postures available, from normal repose to actual flight or attack. These preparatory actions—often incomplete, appearing at the beginning of some activity before motivation is at its peak or sharply focused—inform the animal's social partners, and also inform us, of what it intends to do. They are *movements of intention.* Many of these movements, variable in intensity of expression, have undergone evolution, which has selected the most conspicuous forms by reason of their value as signals and their usefulness in communication. Numerous movements of intention to take flight, performed conspicuously and unequivocally, often displaying certain structural characteristics—for example, outspread wings revealing a white rump—serve to alert other members of the group of imminent danger. Many threat displays, fixed in form and intensity, are also derived from intention movements of flight or attack.

A behavior pattern may be related to not just one, but to a combination of emotions. We know that motivational conflicts result in ambivalent behavior patterns, substitution, or redirection activities. If, for example, the attack and flight drives are activated simultaneously—and that often happens when an intruder comes into view—the result is simultaneous ambivalence, which produces a compromise movement with clearly recognizable components of attack and flight. These two elements are present in varying proportions according to the strength of the stimulation which activated each of the two drives.

In laboratory research, the two centers responsible for attack and flight can be stimulated to varying degrees, either by simultaneous stimulation by means of electrodes implanted in the two corresponding areas of the brain, or by the presentation of models bearing various combinations of the stimuli which release attack and flight. Thus, the experimenter can obtain the full spectrum of all the compromise gradations possible between the extreme responses of attack and flight. At least that is the case in species considered by morphological criteria to be most primitive. But in species considered to be more highly developed, again on the basis of morphological criteria, it is found that a wide range of compromises no longer exists. There remains only one single combination having a fixed value. It is no longer possible to modify the relative proportions of attack and flight in compromise movements by manipulating the stimulus situation, by activating the attack or flight drives more or less strongly. Evolution has, in effect, selected and fixed a constant combination which has particular value as a signal. Huxley, in 1914, half a century ahead of his time, called this process *ritualization*. Baerends and Blokzijl (1963), studying movements of compromise between attack and flight, found that fishes of the *Tilapia* family express this compromise by spreading their fins to a greater or lesser degree, and hence they were able to show that in a particular movement the attack/flight ratio determined by evolution can be different in different species.

A comparative study of inciting movements in shelducks and mallards (Lorenz, 1941) is an excellent concrete illustration of the preceding hypothetical example involving more primitive and more developed species. In a shelduck pair, the more primitive species, the female is the more active partner in threatening and chasing off intruders, and she stretches her neck and shakes her head in their direction, inciting her mate to join in. The following scenario unrolls: neck extended horizontally, the female rushes forward in headlong attack; but the closer she comes to the enemy, the more her drive to attack is counterbalanced, and finally outweighed, by the drive to flee; she does an abrupt about-face, and runs for refuge toward the male; but as she approaches him, the flight drive dwindles and attack regains the upper hand; with her body still turned toward her mate, the female stretches her neck toward the intruder and threatens him afresh. This expression of behavior, incitation, is not an inherited action pattern, but is a compromise solution between two variable pressures, the two activated drives. Furthermore, the ultimate orientation of the female's neck is determined by the relative location of the three protagonists—male, female, and intruder. Consequently, the axis of the female's body and that of her neck can form any angle.

Only one of the many final positions possible in the female shelduck—a position with the body turned toward the male and the neck thrust back over the shoulder—is observed in the mallard, the more evolved species. A female mallard does not take the location of the participants into account, but thrusts her head back over her shoulder even if in doing so her bill is pointed in the direction opposite to that of the intruder. Moreover, the mallard's incitation is not expressed as a variable compromise between attacking and fleeing. It is performed as an indivisible whole, a set of coordinated movements welded together into one block. Thus a movement, which in the shelduck is a compromise resulting from conflict, becomes in the mallard a new, fixed action pattern. It should be pointed out, however, that the mallard's thrust of the head over the shoulder still includes an orientation response. If the intruder is in front of the female, she keeps her eyes on him and does not move her head back very far. If, on the contrary, the intruder is to her rear, the female darts more quickly and thrusts her head further back over her shoulder. The orientation response and the new, fixed action pattern are superimposed.

Two rival black-headed gulls (Tinbergen, 1959), males confronting each other, will hold the neck and head either more or less extended and horizontal, or more or less upright and vertical, depending on whether attack or flight is predominant in the motivational conflict. Two courting gulls, male and female side by side, will invariably extend the neck horizontally while keeping the head upright. Thus, an ambivalent movement, reflecting the attack-flight motivational conflict of a combat situation, has become ritualized, and it is employed in a fixed and nonhostile form in courtship display.

When two instinct systems are activated simultaneously with equal intensity, they act as a brake upon each other, mutually inhibiting each other, and the motivational energy which cannot be discharged through either of these two blocked routes finds another outlet offering less resistance, thus causing unusual expressions of behavior—substitution activities. The prime function of such activities is to discharge an energy surplus which cannot proceed through the normal channels. But a substitution activity can take on a second role. It can be ritualized, that is, modified, adapted to a particular situation, and in the process it acquires a new meaning. Preening of feathers in ducks is one example. All ducks have a behavior pattern of personal grooming which consists of preening or polishing the wing feathers. This behavior may be used as a substitution activity in certain conflict situations. The substitution preening may, in turn, be integrated into courtship, and there, through the process of ritualization, preening becomes adapted to a new function as a social releaser.

Comparative ethology shows that in certain more primitive species of ducks, courtship preening is scarcely distinguishable from personal grooming: in both instances, the shelduck works over the wing feathers with its bill. In other species, courtship preening differs more markedly from bodily grooming and can be more easily recognized by its distinct form: the mallard raises its wing to reveal a white patch beneath, which it further emphasizes by vigorously riffling through the feathers; the garganey raises its wing and with its bill points up the blue wing feathers. In the most advanced species, courtship preening is extremely different from the grooming movement, and its origin can be determined only by tracing the evolution of preening in other species: the mandarin duck uses a very precise bill movement to point out a bright orange patch of plumage. By comparing the degree to which this movement has been ritualized in different species of ducks, it can be seen how preening, a simple substitution activity to begin with, acquires a new function when it is integrated into courtship behavior, and how ritualized preening, in adapting to particular situations and emphasizing specific colorations, departs further and further in form from original grooming behavior. In this example, ritualization consists of exaggerating a natural commonplace movement, transforming it into a signal that is conspicuous, unambiguous, and precise, and integrating this signal into courtship display to serve a communicative function between social partners.

Cranes also preen their feathers as part of their normal grooming behavior. This movement has become ritualized to varying degrees in different species, and it has also been adapted to a new function, but in this case as a threat gesture. In certain species, threat preening remains very similar to the grooming movement. In other species, quite the contrary, it differs markedly and has all the signaling characteristics—clarity, precision, specificity—of a social releaser.

The ethogram of the stickleback (Tinbergen, 1952b, 1953a) provides another well-researched example of the ritualization of a substitution activity. After fertilizing the eggs, a male stickleback takes up a position at the entrance to the nest, and there he circulates the water by a fanning motion of the pectoral fins, thus aerating the eggs. In experiments, this fanning movement can be activated by increasing the concentration of carbon dioxide in the water. A stickleback may also perform this fanning at other times than during the parental phase of the reproductive cycle. It can often be observed during territorial defense, when excess sexual energy is discharged in this way. At this time, the fanning is a substitution activity which has no communicative function, but simply serves to discharge impulses blocked elsewhere. In courtship, however, when the male places himself before the nest to induce the female to enter, he executes a movement which exactly resembles parental or substitution fanning, and this is a case of ritualized fanning, with the function of pointing out the entrance to the nest where the eggs are to be laid.

If the behavior of the stickleback is compared to that of a closely related but more primitive species, it is found that in the latter species courtship fanning has not been so highly ritualized. Comparing homologies—the same method used to study preening in ducks and cranes—makes it possible to retrace the process of ritualization of the stickleback's fanning. It should be emphasized that courtship fanning and parental or substitution fanning depend on different motivations. The ritualization process implies a restructuring of behavioral mechanisms. A ritualized behavior pattern, the stickleback's fanning as well as the mallard's incitation, is not activated in the same way as the original movement.

In their social relations—in agonistic, sexual, or parental behavior—animals are frequently subject to motivational conflicts. Hence, it is not surprising that animal courtship displays consist of sequences or combinations of ritualized intention movements, redirection activities, substitution activities, etc. Courtship behavior has been most thoroughly studied in birds, whose displays are often quite spectacular. Already at the beginning of the century, ornithologists, fascinated by these exhibitions, had described them in all their minute and complex details. But the origin of bird displays remained a mystery.

During the course of its reproductive cycle, a bird performs a series of natural and ordinary movements in response to motivation and specific external stimuli. Activities such as gathering nesting material, passing this to the mate, providing food for the incubating female, feeding the young, etc., normally occur in well-ordered succession. However, these behavior patterns may also appear prematurely, not in their normal context, for example when the birds are first pairing off. At such times, these behavior patterns have no connection with their original goals. They are no longer dependent on the drives to build, to feed, etc. They are registered in quite another context—courtship display—and they have a new function. They are expressed in a more ostentatious and stereotyped form. In short, they have become a veritable *ritual*. It was in observing the courtship display of the great crested grebe, in fact, that Huxley realized the evolutionary import of this behavioral transformation, and he called it *ritualization*.

Two grebes dive under the water, then surface, each bearing a clump of algae in its bill; they approach each other, come face to face, then rise up out of the water breast to breast (Figure 23). In a pair of greylag geese, each partner plunges its head in the water and then draws it back toward its side; this synchronized move-

ment is repeated rhythmically and precedes mating.
These movements are actually rather stylized forms of
movements normally used to gather nesting material.
A pair of courting marsh hawks approach each other
in flight; one of them somersaults onto its back and
they touch talons. This ceremony recalls—or rather
heralds—the exchange of food between male and female
which will occur at nesting time. At that time, a male
announces his return from the hunt by whistling to the
female in the nest; she takes flight to meet him; when
they come near, the male drops the prey, and the female,
rolling over on her back, catches it in her talons. A male
tern offers his prospective partner a little fish he has
caught. This transfer of food seals the betrothal. A cormorant, on the other hand, presents his intended with a
twig or a feather. An exchange of food or nesting material is a frequent feature in the courtship display of birds.

FIGURE 23. Courtship of the great crested grebe: the two birds, face to face, rise
up out of the water bearing clumps of algae in their bills. (After Huxley, 1914,
1966.)

Although ornithologists had described these courtship
ceremonies down to the last detail and had accurately
conveyed their ritual nature, it remained for Lorenzian

ethologists to clear up the question of their origins. They have shown that the collection of nesting material and exchange of food observed in bird displays were initially substitution activities, but they have been transformed and have become essential elements of these courtship ceremonies.

A ritualized behavior pattern, then, evolves from some ordinary behavior, a fixed action pattern, which later loses its original function and, on being integrated into a new context acquires a new communicative function. It becomes a signal which should elicit an appropriate response in a social partner, without delay, by a sort of ethological reflex. Moreover, it can assure a discharge or reduction of aggressiveness in an innocuous way, for instance through the war games which make up territorial combat. When repeated in lengthy ceremonies, ritualized behavior also ensures coordination and synchronization between partners; it strengthens and secures the sexual bond. Actually, when two partners approach each other, especially at mating time, aggressive tension mounts. Aggressiveness is rerouted and channeled through the expedient bypass of ritualized greeting ceremonies, synchronized performances which serve to reassure the pair. This kind of ritualized behavior is seen in the bill-clapping of swans, the head-tossing of cormorants each time they return to the nest, the triumph ceremony of geese, etc.

The results of research on ritualization were compiled at a symposium presided over by Huxley in 1966, some fifty years after his first inspired insight. From all the studies which have been made of this type of behavior, there emerges a concept of ritualization which implies three conditions.

The first is *crystallization*, fixation of a movement so that subsequently it is always performed in an identical manner. Its form, amplitude, velocity are all stabilized and become constant. This is the concept of *fixed or typical intensity* (Morris, 1957). In order to be effective and understood, a social signal must be clear and unequivocal, impossible to mistake. The intensity of a ritualized response does not vary with a weaker or stronger stimulus; only the duration is affected. A ritualized movement—constant in form, amplitude, and velocity—is *repeated rhythmically* for a variable length of time.

Second, a ritualized movement undergoes *changes* by which it comes to differ from the original movement. In adapting to a new signalling function, it becomes more ostentatious, exaggerating dramatic effects. Along with these changes, new shapes, structures, colors are accentuated and developed. Structures and movements underscore one another to enhance the signaling value of the behavior pattern still further.

Finally, there is a *restructuring of mechanisms*. Ritualized behavior, a new fixed action pattern, becomes a discrete innate behavior pattern, dissociated from its prototype and independent from the original motivational conflict. It is integrated not only into a new context, but into a new causal system. It acquires "the full autonomy of an independent instinct," having its own innate releasing mechanism, its own motivation, and its own appetitive behavior (Lorenz, in Huxley, 1966). This is the *concept of emancipation* (Lorenz, 1961).

Experiments must be delicately balanced in order to prove that a behavior pattern has become independent from the original mechanism—vis-à-vis the original normal action pattern and also vis-à-vis the same movement used as a substitution activity. Experimental proof can be obtained by showing that conditions which are increasingly favorable to the expression of the original movement do not increase the probability of emission of the ritualized response. It is known that an increase of carbon dioxide activates parental fanning in the stickleback. Substitution fanning, indicating excess sexual excitation during territorial defense, can also be encouraged by an increase of CO^2, since motivational impulses are more easily discharged through an alternate channel if the external stimuli which unblock that route are present. (Two rival cocks perform substitution pecking more readily if grain is tossed on the ground.) However, if emancipation has been achieved, an increase of CO^2 should have no effect on the frequency of ritualized fanning which has been integrated into the stickleback's courtship.

Behavior and Speciation

Although inherited action patterns are stable and resistant to change, and although they testify to the unity and homogeneity of a taxonomic group, the behavior of closely related species can nevertheless differ at various levels and can diverge in several ways. Action patterns can undergo a microevolution in form. More frequently, differences in response threshold or frequency of use develop. These quantitative variations may ultimately result in the loss of a fixed action pattern. Thus, in the final analysis, a quantitative change in a behavior pattern may be translated into a qualitative change in the behavioral make-up as a whole. Related species can combine and order commonly inherited action patterns into species-typical sequences, so that each sequence functions as a signal which is understood by a particular species. Emotionally motivated action patterns, singly or in combination, can become social releasers; they can be crystallized, emancipated, and adapted to particular situations, through ritualization.

It can be seen that once an action pattern has become a social releaser, and once it has reached a certain degree of specificity, it acts as an *isolating mechanism,* and it strengthens—or guarantees—speciation. Ritualized behavior clearly identifies a species and prevents crossbreeding. A case in point is courtship preening, which is modified and ritualized to varying degrees in different species of surface-feeding ducks. In each species, the movement of the bill is different, pointing out and emphasizing certain brightly colored plumage. Another instance can be seen in the betrothal ceremonies of these ducks, all composed of the same elements combined in different ways. Male fruitflies court the females with wing vibrations, but these movements vary from one species to another, another case of homologous movements which originate from a fixed action pattern common to all and then diverge. A male fiddler crab (cf. Crane, 1957 and in Huxley, 1966) signals by waving its huge claw, prominently displaying the unusual structure and coloration. This claw, and these movements, too, differ from species to species.

When two populations have been separated, divergent evolution during this period of isolation not only leaves its mark on the structural level, but also puts its stamp on ecological preferences and behavior. Marlier (1959) studied a small cichlid fish, the *Tropheus mooeri,* which is distributed throughout Lake Tanganyika. Local populations are very similar on the morphological level, but they differ from one another in their courtship coloration. It would be extremely interesting to determine by controlled experiments to what extent these color patterns have become social releasers, what role they play in courtship, whether they serve as isolating mechanisms and guarantee group cohesion.

Competition exists where closely related species live sympatrically. This competitive situation acts as a selection pressure to accentuate existing differences, and they in turn serve as isolating mechanisms and minimize the competition. In regions of western Europe where three species of warblers, very similar morphologically, live side by side, each species shows a decided preference for a particular habitat: the wood warbler prefers the tall timber, the chiffchaff selects forest underbrush, and the willow warbler chooses thickets of small trees. But in an area where only one species is present, its ecological tastes are broader. In this case, there is no competition, no risk of hybridization, and hence no selection pressures for isolation. Bird song is used as a territorial signal addressed either to females or rival males. The songs of related species have evolved from calls, that is, muscular contractions of the syrinx, which were originally very similar. But they have become ritualized and often differ markedly in closely related species. Nevertheless, in an area where only one species is present, these differences are muted, and the song becomes less specific, since the selection pressure favoring specificity is lacking. This occurs in the song of chiffchaffs which have no competition from willow warblers (Thielcke, in Huxley, 1966). Another example is the song of meadow pipits in Iceland where there are no tree pipits. In the same vein, sexual dimorphism is very pronounced in ducks if several closely related species share the same wintering grounds. That is where the bethrothal ceremonies take place, and each of the females chooses her companion for the coming year from among all the males decked out in their courtship finery. On the other hand, in aleopatric zones where only one species is found, for example on islands far at sea, and where there is no risk of confusion in the choice of a partner, dimorphism is blurred. The male tends to acquire plumage quite as drab as the female's, and even his sexual behavior tends to be less elaborate (Johnsgard, 1965).

The absence of selection pressures, then, reduces the specificity of social releasers which act as isolating mechanisms. An experiment on fruitflies (cited by Tinbergen, 1966) demonstrated the corollary of this principle, that is, that the presence of a selection pressure—in this case artificial selection—can favor existing behavioral differences, increasing their specificity and thus encouraging isolation. The experiment started with a breeding stock of two varieties of fruitflies, each variety showing only a slight preference for its own kind, and, as is usual, they crossbred quite readily. All hybrids of the first generation were eliminated, and only the two pure strains were preserved. The pure descendents of the two groups were mixed, and their hybrid offspring were in turn eliminated. This process was continued through forty generations, stretching over a period of three years. In the forty-first generation, the males showed a very marked preference for females of their own variety, and the females, for their part, when courted by males of the other variety, scarcely responded at all. Artificial selection acted on the original variation, favoring the preference of a variety for its own kind, favoring isolation of the two groups. This experiment shows very clearly how natural selection favors and guarantees separation and speciation by acting on the level of behavioral variability.

CHAPTER 5

THE DEVELOPMENT OF BEHAVIOR

Instinct and Learning

In the preceding chapter we have seen that an animal inherits a behavioral framework consisting of behavior patterns and mechanisms which are adjusted to the environment and way of life peculiar to the species. Adjustment is accomplished gradually, from generation to generation, through the action of selection pressures which favor those individual members of the species whose behavioral make-up is best suited to the environment. This adaptation on the level of the species is characterized by an evolution of behavior patterns during the history of the species. On the other hand, there exists a process of adaptation which occurs on the level of the individual, and this is characterized by changes which take place during the history of the individual. In certain circumstances, individual behavior can change and develop, and it can be affected by acquired experience.

An animal, then, may be adapted to the environment by receiving an adequate behavioral endowment to begin with. An animal may also be adapted to the environment by shaping its behavior progressively, according to the conditions it encounters. The relative importance of these two processes of adaptation varies, depending on the behavior patterns and species in question. Clearly, an

arthropod, an animal whose normal life span is sometimes limited to a few weeks, an animal that never knows
its parents nor experiences a family environment, must
be equipped to function immediately in an adequate
manner. This is less essential for animals raised within a
family circle, animals which develop more slowly and
have a longer life span; they have an opportunity to adjust and perfect their behavior in the course of their
upbringing and maturation.

It can be stated as a general rule that the ultimate
behavior of an individual animal is, in varying degrees,
the result of a complementary interrelationship between
inherited behavior, which is determined on the same
basis as physical attributes, and learned behavior, which
is adapted to the circumstances. The black-headed gull
provides an excellent illustration of this complementary
interrelationship. The progressive development of a
chick's behavior shows an increasingly complex combination of innate and acquired elements (Tinbergen, 1966).

The chick begins by breaking through the eggshell
with its "egg tooth," a small protuberance on the top of
the upper mandible. This is an extremely important action which is performed only once in a lifetime, but the
chick performs it straight off in a satisfactory manner.
Very soon, even before its down has dried, the chick
starts pecking at the parent's bill, and it begins to take
food and swallow it. During the first day, the chick
stands on its legs, takes its first steps, preens its plumage.
If the parent calls in alarm, the chick crouches down. In
two or three days, it reacts to this same call by first running away and then crouching. Later, it runs away and
hides before crouching. Still later, it runs away, hides in
a chosen place that it knows, and then crouches. At one
week, the chick makes its first flying movements. At two
weeks it calls its fellows. At three weeks it feeds indepen-

dently. At first the chick pecks indiscriminately at various objects, but later it becomes more selective and pecks only at what is edible. At five weeks, the chick makes its first trial flights. Initially, it alights clumsily and somersaults on landing, but it soon learns to land into the wind. At first the chick heads toward any glittering surface in order to drink, and only later learns to distinguish water. It first goes through all the motions of bathing while on land, and subsequently learns to perform them in the water.

Certain movements are perfect right off, while others are perfected progressively. Certain movements are performed deliberately from the outset. Others are seemingly aimless at first, and are later adapted to the circumstances, adjusted to the situation.

For generations, researchers have argued over what is innate and what is acquired in behavior (cf. Lehrman, 1953; Lorenz, 1965). The distinction is indeed difficult at times, and mistakes have been made, which is inevitable unless an animal's behavioral development is studied from its birth, under natural conditions as well as under experimental conditions. A behavior pattern cannot be termed innate just because it is stereotyped and identical in various individuals of a species. It is normal for members of the same species, placed in identical circumstances—and that is usually the case when their habitats are identical—to tend to learn the same thing and to express it in the same way. Nor can a behavior pattern be termed acquired just because it is not performed from the outset in its perfected and definitive form. Actually, this dispute is rather hypothetical and academic, for the complementary interrelationship between the innate and the acquired is dynamic and continually in a state of flux.

Inherited behavior falls into place step by step. Certain

responses are expressed very early in life, such as the reaction to flee from predators, and movements related to physical comfort and body care. Others do not appear until much later, when the individual has become an adult, notably all the behavior patterns linked to reproduction, for example the territorial combat of vertebrates. This gradual emergence of innate behavior is accompanied by the appearance of learning processes, and in fact certain learned behavior may be manifested before innate behavior has fully matured. Furthermore, not only is there simultaneous development of the learning process and innate behavior, but a close interdependence exists between the two. Genes determine potentials, their realization being dependent on environmental conditions. Or stated conversely, and more specifically, what an animal can learn is fixed by its genetic make-up, is predetermined, so to speak, by its innate equipment. As a matter of fact, animals inherit, as part of their behavior machinery, dispositions to learn certain things at certain times.

An animal's ultimate behavior is a function of genetic and environmental input acting in concert. The development of behavior, innate as well as acquired, depends first of all on the physical and functional maturation of the internal machinery which underlies behavior—skeletal, muscular, and nervous systems; physiological mechanisms; sense organs; etc. And it is from this angle of structural maturation that we should first examine the development of individual behavior.

Emergence and Maturation of Behavior

An embryo is capable of performing ever more complex and perfect movements as its structures develop. A tadpole makes swimming movements while it is still within the gelatinous matrix of the egg. These movements improve from day to day, at first during the embryonic stage and later in the freedom of the larval stage. Carmichael has shown, however, that this improvement is not due to practice. He placed a group of eggs in a water solution containing a light anesthetic, so swimming movements were precluded but structural development was unhindered. When the embryos were later placed in pure water, they performed swimming movements which were from the outset as well developed as those of control embryos which had not been anesthetized. Hence, the swimming movements had not been learned. Rather, at any specific time, an embryo performs perfectly those movements which are possible given the existing neuromuscular development and coordination.

In a salamander embryo, muscle segments and neuromuscular connections develop and become functional progressively, proceeding from head to tail. If a negative stimulus is applied to the right side of the head of a young embryo, its first response is to move the head in the opposite direction, by a reflex contraction of the first muscle segments on the left side. With time, more segments begin to function. The same stimulus then produces a wave of contraction which flows from head to tail, and the trunk of the animal curves in to form a circle. Still later, neuromotor connections are established, so that sensory nerves are linked to muscle segments on the same side. When a stimulus is then applied to the right side, contractions first appear in the left segments and the front part of the animal begins to curve in, but by the time the contraction wave reaches the tail, contractions begin to occur in the right segments, again proceeding from head to tail. Thus the embryo does not curve into a circle, but rather into an S-shape, and its escape response takes the form of an undulating swimming movement. The appearance of this swimming movement is not related to a learning process. It is linked to the gradual maturing of the motor and nervous systems.

There are various other methods of proving that learning and practice are not responsible for the development of such behavior. If a tadpole's entire spinal cord is isolated from the afferent nerves, thus preventing sensory input from affecting its movements, coordinated swimming movements develop nevertheless. Similarly, an animal can be prevented from making learning movements by confining it in a tube; when released, it can perform perfectly the same complex movements as unrestricted control animals. This last experiment parallels the conditions encountered by birds raised in narrow cavities. After having passed several weeks confined within a rocky fissure or restricted to a narrow ledge, young black martins, without any preliminary practice, must launch into space and take flight. In the same way, young sea birds which are raised in the niches of steep coastal cliffs must, unrehearsed, dive straight into the sea.

Weiss operated on salamanders before neuromuscular connections had been established. In one experiment, he interchanged the antagonistic muscles of one leg, and in another, he exchanged the right and left anterior limb buds. When the connections were established, both the muscles and the limb buds moved exactly as they would have done in their original positions, showing no adaptation whatsoever to their new position. The anterior-posterior axis of a limb bud is predetermined, so the transplant of the second experiment resulted in the limb buds being turned toward the rear rather than toward the front. When a negative stimulus was applied to the snout of the animal, the animal's movement to escape was translated into forward motion. For a year, the salamander proved to be incapable of adapting to this situation. Learning was powerless to change the innate action pattern. The animal was at the mercy of its neuromuscular blueprints.

An analogous experiment can be made with a frog by grafting a piece of light-colored belly skin onto the back and grafting a piece of dark-colored back skin onto the belly. Each of these grafts develops normally, that is, in spite of reversed locations, they maintain their original texture and color. After the nerves have become connected with the skin, the frog can be induced to scratch its back by stimulating the graft on its belly, and, conversely, belly scratching results from stimulation of the graft on the back. Belly skin, even when grafted on the back, relays the message, "scratch belly." The skin supplies information about its nature, not its location, and the frog in this experiment, like the salamander in the preceding one, never learned to adjust its behavior.

These experiments provide evidence of preprogramming, for once connections are established, the salamander and the frog both execute movements in a predetermined form and direction. These movements are aberrant in the circumstances of the experiments, but habituation and learning can do nothing to change them. One must be careful not to exaggerate the influence of learning or the importance of practice in the maturation of behavior. The majority of motor patterns are extremely resistant to change, and even in man, adaptation, re-education, and readjustment occur infrequently and very gradually.

I shall cite one more example—one which anyone can easily verify—showing that certain behavior cannot be swayed by learning. At the biological research center at Zwin, and in many parks and zoos, there are ponds teeming with numerous species of ducks living in semi-liberty. Some of these are wild ducks which are attracted to the ponds by the ducks already there. The latter ducks were originally wild, but have had their wings clipped. This operation is performed on one wing and consists of cutting off the tip of the bone bearing the primary quills. A clipped bird can no longer fly. At the very most, it can rise only a few inches before falling back into the water. The ducks in the ponds, clipped or not, become accustomed to visitors passing by, and their tendency to flee when approached diminishes. But when they are disturbed by an unaccustomed intruder—a dog or a visitor crossing over a barrier—the ducks make movements inciting to flight. When anxiety reaches a certain level, and when incitation to flight reaches a certain intensity, the birds take off. The wild ducks, the control birds in this instance, fly away. The clipped ducks, however, lose their balance, pivot to one side, and somersault right back into the water, or even fall onto the cement around the ponds. In spite of repeated punishment, these ducks never learn not to take off.

Inherited action patterns find expression as their supporting structures become functional. The ontogenesis of behavior goes hand in hand with the growth of the embryo and the development of its organs. There is a timetable for the appearance of behavior patterns which àre directly dependent on structures. There is also a timetable for the appearance of behavior patterns which are more highly integrated than simple motor movements. The various elements which go to make up a complex behavior pattern often reach maturity independently, and they are later combined into sequences of a higher order, into sequences which are functional, meaningful, and goal-directed.

When only two or three days old and still devoid of any feathers, a young stork nestling performs a movement putting its head over on its back. This fixed action pattern will not be integrated into a sequence until very much later. A young gull first executes a number of separate and distinct movements in which it extends its wings and its legs. Later it will combine these movements into a wing-cleaning pattern. An adult cormorant completes its nest by inserting twigs into the nest wall, and, in the process of forcing them in, the cormorant shakes its entire body. This trembling movement first appears very early in young birds, without rhyme or reason; only much later is it integrated into a purposeful context.

In ontogenesis, then, single units of relatively simple behavior, fixed action patterns, appear first, and more generalized behavior and ordered sequences appear subsequently. Simple motor patterns are expressed early on, without object, without reason, and independently from one another. Later they are combined with one another and with appetitive actions, into functional and goal-directed sequences.

The behavior repertoire is only completed with adulthood, with the advent of sexuality and reproduction. Sexual behavior is linked to the maturation of the endocrine system. In fact, sexuality can be induced prematurely by the injection of appropriate hormones in proper amounts. This phase is the final one in the maturation of innate behavior. From then on, adult behavior will continue to manifest seasonal variations—migration, reproduction, etc.—in accordance with a rhythm regulated both by genetically determined factors and by environmental factors, the relative importance of these factors varying with the species. Finally, variations will occur within each reproductive season, for the order in which specific behavior patterns appear and the complex way in which they develop are determined by a subtle combination of internal and environmental factors (see Chapter 3).

Interaction between Fixed Action Patterns and Learning

We have examined how the development of behavior parallels, and is indeed dependent on, the development of the physiological machinery of behavior. Innate behavior emerges as the individual's musculature, neural circuitry, and endocrine system become fully functional. There is a wealth of evidence, however, that behavior patterns, as observed in their final form, result not only from the maturation of organic substructures, but also from interaction between inherited elements and a learning process. There are some very simple and obvious illustrations of this interaction. In order to open a nut, a squirrel uses three inherited movements: handling, gnawing, and prying open. A young squirrel knows how to perform these three operations from the outset, but it must learn to combine them effectively. Only after some time does a squirrel learn to manipulate a nut in such a way that it bites the shell at the point of least resistance and breaks it open cleanly and quickly. The remains of nuts which have been tackled by young and inexperienced squirrels bear witness to their arduous early efforts.

There are also very complex examples of interaction, for instance, the songs of certain mimic passerines such as nightingales, larks, finches, and bush warblers. The songs of these birds are combinations of inherited motifs, common to all the members of a species, and learned elements, acquired by imitating either their own species or other species. The marsh warbler's song, or that of the shrike, is a veritable potpourri made up of typical motifs of all the species which frequent the mimic's habitat. This eclecticism is well known, and the fowlers in the region of Liège have put it to practical use. They breed and raise birds to be used as decoys or entered in, competitions, particularly canary-goldfinch hybrids. Traditionally, these young birds are placed at a very early age under the tutelage of a "professor," a bird selected for the beauty of its song.

The song of these passerines is prime material for studying the development of behavior. Bird song consists of a typical sequence of muscular contractions translated into sounds. Today, these sounds can be recorded with great fidelity and retranscribed in a sound spectograph to obtain a graphic representation of the contractions of the syrinx. Sound spectograms make it possible to analyze individual songs, to compare the songs of birds raised in isolation with the songs of birds raised under various other conditions, and thus to differentiate inherited patterns from learned patterns with precision and certainty. It is paradoxical that of all the various types of motor patterns, bird song is today the most amenable to rigorous analysis. Although it had been treated extensively in specialized works, until a decade ago it was one of the aspects of animal behavior least susceptible to objective study and experimentation. One was forced to devise various methods of musical annotation, quite ingenious, but not very thorough and inevitably subjective. Recent advances in electronics paved the way for the outstanding work which has now been achieved in this field.

Marler and Tamura (1964) discovered that American white-crowned sparrows have local dialects. Young birds, which had different regional origins but had been raised in isolation, produced a basic song which was always identical, common to all. This song is a fixed action pattern, characteristic of the species. Regional dialects come about through learning. This process takes place during a young bird's first three months of life, long before the bird itself ever sings. Its first song will not appear until the following spring when the bird, now adult, establishes a nesting territory. If a young male is captured in its first fall, that is, after three months of learning, and is subsequently kept in isolation, the following spring it will begin to sing in the exact dialect of its native region. Up to the age of three months, any white-crowned sparrow can be taught to superimpose any dialect on the basic innate song by playing recordings to it. If the young bird is kept in isolation during its first three months and then exposed to recordings in the fourth month, the experiment will have no effect, for the learning period has passed. This bird will know only the basic song which it inherited. In all cases, the results of the experiments cannot be assessed until the following spring.

Konishi (1965) pursued the analysis still further. He selected white-crowned sparrows of varying ages and made them deaf by removing an ossicle of the inner ear. The birds could hear neither the sounds of those around them nor the sounds which they themselves produced. Konishi found that if the operation is made very early, the following spring the bird will produce a succession of disconnected sounds in which the basic song cannot be recognized. It follows that the white-crowned sparrow must hear itself in order to express its innate song. If the operation is performed after a young bird has heard a complete and normal song during the learning period, but before it has ever sung itself, this bird, too, will be incapable of uttering any song but one made up of disconnected sounds. Hence, the sparrow must hear itself not only to express the song it has inherited, but also to reproduce sounds it has learned. Finally, if the operation is carried out in the second year of life, after a bird has sung correctly at least once, the operation proves to have no effect and the song remains properly phrased and recognizable. This will be the basic song if the subject was isolated during the learning period; it will be the complete and perfect song if the subject had the opportunity to learn. Bird song has been most thoroughly researched, and it is the best illustration we have of the complementary relationship and subtle interaction between heredity and learning. This type of study should be extended to other behavior patterns as well.

A difference of behavior is usually thought to be an indication of learning, while identity of behavior in various individuals is taken to be a sign of innateness. Since learning depends on individual experience, it may be assumed that individuals will not learn the same thing if placed in differing circumstances. This holds true in the laboratory, but in natural circumstances, members of the same species are very likely to learn the same thing, since they usually live under identical conditions in identical habitats. A nightingale learns the species song through contact with its parents. It would be impossible to ascertain that its song requires a measure of learning, except for the fact that nightingales have been raised in isolation and among other species. In order to determine accurately what an animal really inherits or learns, the animal must be raised under varying conditions and its resultant behavior compared with normal behavior. Raising an animal in isolation is essential in order to distinguish innate behavior, but it is also important to diversify the conditions under which an animal is raised, since the actual expression of innate behavior potentials is dependent on environmental conditions. This method of investigation is useful in determining what factors influence the development and expression of behavior patterns and to what degree, what factors cause delays, deviations, omissions, or modifications.

Innate Disposition to Learn

The naturalist's greatest interest lies not so much in finding out what an animal is capable of learning, but in discovering what the animal actually does learn and determining how advantageous this learning is. What an animal can learn is fixed by its innate machinery, and, by the same token, what it does learn is often determined by heredity. Dispositions to learn may vary from one species to another and in accordance with the behavior in question. Once more, bird song provides a good example. The American white-crowned sparrow is capable of learning any of the regional dialects of the species, but it is incapable of learning the song of a different species. The European warbler finch (Thorpe, 1961) also possesses regional dialects, and its song is composed of a basic inherited pattern to which learned motifs are added. This finch is best equipped to learn local variations of the song of its own species, but it can learn the songs of other species, although it does not do so readily. The song it learns most easily is that of the tree pipit, and this song happens to resemble its own. This would indicate that the warbler finch's song-learning is preprogrammed with fixed limitations. The green finch and the bullfinch are closely related to the warbler finch. All belong to the Fringillidae family, and all are equipped with essentially the same vocal organs. At first, the green finch and bullfinch do not sing as well as the warbler finch, but they are much better imitators and can learn a wide variety of songs. Important learning differences do exist, not necessarily in capacities, but in the built-in range of possible variations, the range being wider or narrower according to the species. It is not very clear what interpretation should be given to these differences. There are, however, dispositions to learn which are clearly advantageous and adapted.

A herring gull (cf. Tinbergen, 1953b) never learns to recognize its own eggs from those of other pairs, although the coloration and markings are sometimes quite different. It recognizes an egg as such—and proceeds to incubate it—by a specific set of characteristics (size, spots, contrast) which can vary within certain limits. But it does not differentiate within these variations. The incubating gull is not at all perturbed if its entire clutch of eggs is switched with another, replaced by artificial models, etc. (see Chapter 2). For several days after hatching, a herring gull is equally incapable of recognizing its own chicks. Little strangers can be substituted and will be readily accepted. But after five days of living together, the herring gull will accept no further substitutions; it recognizes its own young, and it distinguishes them from those of other pairs. To our eyes, however, these chicks all look alike, whereas the eggs were easily recognizable. Why does the gull behave differently toward the chicks than it did toward the eggs? It is very important for the herring gull to learn quickly to distinguish its own chicks, because after a few days they begin to move about and intermingle with the neighboring chicks. On the contrary, there is nothing to be gained from learning to recognize its own eggs, because there is no likelihood that these will go for a stroll outside the nest. During incubation, recognizing eggs is of no consequence. What is important at that time is finding and recognizing the nest. As a matter of fact, if the experimenter removes the eggs but places them right next to the nest, the herring gull will nevertheless alight and sit on the empty nest!

Other species, living under different conditions, have a real stake in distinguishing their own eggs from those of their neighbors. Many sea birds nest in crowded colonies on the narrow ledges of steep cliffs. These birds often do not build nests, but lay their eggs directly on the bare rock. Their eggs run the risk of rolling about and getting mixed up with those nearby. It is important, then, for each pair to know how to recognize their own eggs. Tschanz (1959) showed that under these conditions guillemots do indeed learn to recognize their eggs. This learning is no doubt facilitated by the appreciable variations of color pattern which occur in the clutches of different females.

Jackdaws, which nest in colonies in trees, show no aptitude to recognize either their own eggs or their own young. A jackdaw colony is organized along very strict lines, in a hierarchical order (see Chapter 6). Each bird has a very definite place in the pecking order, and it comes to know its place by learning. Each individual learns to respect higher-ranking birds for they will peck it, and learns to recognize lower-ranking birds for they will submit to pecks without returning them. In this society, relations between members are regulated by their relative dominance and inferiority, aggressiveness and submission, and all these characteristics are acquired individually. Lorenz studied the jackdaws which lived in semi-liberty at his house, nesting in its gables. He found that when a lower-ranking female mates with a dominant male, she automatically acquires the same rank the male enjoys within the hierarchy. And the whole colony is immediately aware of the promotion. The female, which previously was always being shoved about by one and all, is now socially respected by the entire community. There is a striking contrast between the alacrity with which jackdaws learn to recognize status changes in the social hierarchy and their inability to recognize their own eggs or their own young. They are well equipped with learning capacities in a particular area—the social situation—which is very susceptible to change.

The examples cited, jackdaws, guillemots, herring gulls, and passerines, all inherit dispositions—different in different species—to learn certain things at certain times.

Learning Processes

Thus far I have demonstrated the complementary nature of the relationship between the innate and the acquired. In previous chapters I did not address myself to the task of making inventories of innate behavior, nor shall I now proceed to enumerate lists of learned behavior. That is not the purpose of this book. At this point, however, it would be appropriate to make a brief survey of various learning processes by which individual responses are grafted onto the species-specific inheritance. I have borrowed most of the following definitions from Thorpe (1964), author of a comprehensive work on learning in animals.

Thorpe defines learning as a "process of change, adaptation of individual behavior as a result of experience." The simplest form of learning is *habituation*. This does not involve the acquisition of new responses, but rather the loss of old ones. If a stimulus or a situation occurs repeatedly with neither unfortunate nor advantageous consequences, an animal becomes accustomed to it, and responses which were initially linked to this stimulus disappear—even sparrows cease to be frightened by a scarecrow. Identifying habituation is subject to error, because the disappearance of a response can be due to other causes as well, such as decreased receptivity of the sense organs to repeated stimulation, or simply muscular fatigue, or perhaps a change in motivation. These other processes must be ruled out before one can speak of habituation.

Learning by *association* consists of making a connection between a new stimulus situation and a familiar one, so that the response normally linked to the familiar stimuli becomes associated with the new stimuli. The principle of conditioned reflex, discussed in the first chapter, provides a very simple explanation of a multitude of reactions which we are prone to qualify as "intelligent," especially in animals which are close companions of man. A dog devoted to its master is very attentive to his slightest move or gesture. It makes associations between these actions and relates them to their consequences. The fact of the matter is that what we take for foresight is merely a pure and simple conditioned response, a reaction to some signal which we are not even conscious of giving, but which the dog is well aware of. A dog which trots eagerly to the door is not clairvoyant. It is merely producing a conditioned response to some cue which its master has unconsciously given, to a signal which indicates they are going for a walk.

Lorenz (1968) tells the tale of a parrot which used to call out a very timely "hello" or "goodby" to visitors. The bird achieved its most dramatic effect, however, when it seemed to anticipate a departure, most impolitely, by exclaiming "Goodby!" before the visitor—at least as far as he knew—had shown any intention of leaving. Actually, the parrot had noticed a change in the tone of the conversation, indicating it was about to come to an end. What was taken for anticipation was merely a conditioned response to a specific signal which had been perceived by the parrot. Similarly, all the exhibitions and routines which animals like dolphins perform in menageries and zoos are the result of training which utilizes the principle of conditioned reflex.

I also mentioned learning by *trial and error* and *operant conditioning* in Chapter 1. In these processes an animal first performs actions spontaneously, and then it selects certain operations to repeat or not, depending on whether they have an advantageous effect or not—some effect such as obtaining food, avoiding painful or unpleasant consequences, or, in a more general way, obtaining stimuli which permit the animal to satisfy some drive. This kind of learning occurs as a follow-up to an initiative taken by the animal. When animals grow up in their natural surroundings, trial and error occur frequently during the formative phases of behavioral development, and especially in young animals while exploring or at play.

Latent learning, on the other hand, is the result of an "association of different stimuli or situations which have no immediate advantageous effect." The animal obtains no reward, nor does it immediately satisfy a need, reduce a drive, avoid unpleasantness or punishment. This learning process differs from the preceding one in that what is learned is not evident at the time. It is memorized, stored away to ripen for later use. Latent learning often takes place while an animal is exploring its environment to no particular purpose. In the exploratory phase of appetitive behavior, which is related to a specific motivation and instinct, an animal is goal-directed, the goal being the satisfaction of its drives. There is another type of exploring merely for the sake of exploring, independent from the usual motivations of hunger, thirst, sexual pursuit, maternal care, etc. To be sure, an animal satisfies its need to explore, but the associations it makes during this exploratory phase do not offer any immediate profit.

The maze test provides an illustration to help clarify this principle. If one places a litter of rats at the exit of a maze and places the mother at the entrance, she rushes through the maze seeking her young—or, more accurately, seeking the situation which will allow her to satisfy her maternal drive. Once the mother has found her litter, the experimenter can return her to the starting point and measure the time gain when she repeats the course. One can also put a rat in the maze when it is not impelled by any particular motivation—hunger or thirst, sexual or maternal drives. Nevertheless it will proceed to explore the various paths and become familiar with the layout. This acquired knowledge will enable the rat to run the course more quickly at some later date when it may be advantageous, for example when the rat has been separated from its young.

Finally, *insight learning* is the phenomenon one speaks of when a problem is solved in the wake of sudden understanding, by a kind of inspiration as it were. An animal gives the impression of having at once comprehended the solution, for it adapts its behavior very quickly to the situation in which it is placed, after what seems to be brief reflection and without any preliminary trials. Perhaps the animal makes trials and errors mentally, but such a hypothesis rests on a great many assumptions. The credence given to this type of learning comes from the experiments of Kohler (1927) with chimpanzees. Bananas were placed out of reach of a chimpanzee. Boxes were available which could be stacked on one another, or sticks were on hand which could be fitted together, in order to reach the bananas. The chimpanzee comprehended the operation it had to perform and succeeded in passing the test. But how does this prove insight learning? How does it rule out latent learning? As a matter of fact, the chimpanzee had had the opportunity to play with boxes and sticks previously, so in the experiment it used knowledge which had been acquired in another context. Furthermore, the chimpanzee finally reached the bananas only after trial and error.

Birch investigated the role of acquired experience in the solution of problems by insight. He selected young chimpanzees which had been raised from birth under strictly controlled conditions, and he placed them in the same situation as described above. Only two out of six succeeded in reaching the food by fitting the sticks together, and these two were the only two which had previously had an opportunity to play with sticks. Subsequently, the six subjects were all given sticks to play with, and the test was later repeated. Immediately, all six fitted the sticks together correctly to obtain the food. The experiment established that in these instances the chimpanzees were utilizing knowledge acquired previously in play when latent learning by trial and error had taken place.

These laboratory tests should be considered along with the observations made by Goodall (in de Vore, 1965) in Uganda. She found that chimpanzees probe termite hills by poking twigs into the cracks and openings. After inserting a twig, a chimpanzee will wait a few moments, pull out the stick very gingerly, and then lick off the termites clinging to it. This is not only a matter of using a tool, but the chimpanzee also prepares this tool very carefully. Beginning with bits of vine, stalks of grass, or twigs, the chimpanzee removes any leaves, strips them down, and trims them neatly to obtain smooth, slender stems about ten inches long. When these become blunted with use, the tip is cut off, and when they become too short, the chimpanzee prepares new ones.

Behavior of this order is not reserved exclusively to primates. The California sea otter feeds on mollusks and crustaceans, and it breaks these open by striking them upon a stone. When fishing on a sandy bottom, the otter surfaces from a dive with a stone in one forepaw and a bivalve or a crab in the other. Floating on its back, the otter holds the stone on its chest and then strikes its catch upon the stone to crack open the hard shell. The otter eats the meat, throws away the remains, and dives again—still hanging onto the stone!

Any attempt to explain the process by which such behavior is acquired should be undertaken only with great prudence. Laboratory tests are very valuable in that they make it possible to work with animals whose exact background is known. On the other hand, it is of primary importance to know that behavior patterns on this level of complexity actually do occur in nature. The ideal would be for the experimenter to achieve such a degree of intimacy that he could observe animals in their natural habitat and know the exact background of each and every one of his subjects.

Imprinting

The naturalist-ethologists are best known for their work on instinctive behavior. In the early nineteen-thirties, scientific circles had banished this term from their vocabulary and were totally committed to the cult of conditioning and learning. Ethologists deserve great credit for redefining and rehabilitating "instinct." Hence it may seem paradoxical that it is these same ethologists, labeled "neo-instinctivists," who discovered the most radical form of learning: *imprinting*. The paradox is only superficial, for the discovery of the phenomenon of imprinting was above all a matter of method.

Louis Verlaine (1937), who taught animal psychology at the University of Liège before the war and was the most outspoken advocate of the anti-instinctivist thesis, made elaborate investigations, but due to his methodol-

ogy he never came close to discovering imprinting. He
proposed to find out whether newly hatched ducklings
take to water "by instinct." He placed an equal number
of chicks and ducklings in an enclosure which had a
pond in the center. He found that just as many chicks as
ducklings approached the pond and came into contact
with the water. This contact is unpleasant for chicks,
since their down becomes wet, but it is not unpleasant
for ducklings, since their down has oily secretions and
sheds water. The ducks enjoyed the water and stayed
there, whereas the chicks retreated. Ducklings whose
rump oil glands had been removed also did not care for
their first contact with water and later gave it a wide
berth. Verlaine concluded that the first encounter with
water comes about by chance, and that "the specific behavior patterns related to an aquatic life are, like other
behavior patterns which go by the name of instinct, actually the result of individual experience."

Ethologists approach this problem from quite a
different angle, by putting it back in its logical context.
They know that ducklings are led to water by their
mother. They also know that the first moving object
ducklings see after hatching is their mother. The ducklings follow her as soon as their little legs can hold them,
and thus, tagging along behind their mother, they proceed toward the nearest body of water. However, if upon
first emerging from the shell, ducklings see some moving
object other than their mother, they will follow this object in the very same way. Spalding, in 1873, and Heinroth, in 1911, had already drawn attention to this phenomenon. They found that young precocial birds—chicks,
ducklings, goslings—which have been incubated by a bird
of a different species, follow their adoptive mother, and
those which have been incubated artificially, may manifest filial behavior toward humans who were present at

the time of hatching. In 1935 Lorenz systematized the study of such behavior, he defined the laws which govern it, and he established it on a firm theoretical basis. Lorenz's work in this field provides a brilliant demonstration that there is no need to employ complicated methods in order to discover phenomena and concepts of the greatest importance in biology and psychology—rather, all that is needed are methods having both common sense and sensitivity. Lorenz's first experiments with graylag geese are still the classic illustration of this principle.

Lorenz entrusted the incubation of twenty wild goose eggs to domestic fowl, half of them to a goose and half to a turkey. Near hatching time, he removed the eggs from the care of the turkey and placed them in an artificial incubator. His intention was ultimately to put all of the goslings in charge of the barnyard goose. Although she may not have been able to sit on twenty eggs, she would be quite capable of looking after twenty goslings. Lorenz was present when the first egg hatched in the incubator, and, full of admiration, he watched the gosling go through the various stages of emerging from the shell and awakening to life. Its down scarcely dry, the gosling turned toward Lorenz and made specific graylag contact calls, thus singling him out for its first ceremonial greeting. Lorenz confesses he did not foresee the heavy responsibilities he had unwittingly assumed by all this. While waiting for the other eggs to hatch, he placed the first gosling under the downy breast of the big white goose, but the gosling obstinately refused to stay put. It uttered cries of distress and launched out in pursuit of Lorenz, who was doomed from then on to assume full-time care of the gosling, for it was firmly convinced that Lorenz was its mother. The same thing happened when other goslings were born while Lorenz was present. These younger siblings, however, were not cared for exclusively by Lorenz, a privilege reserved to the first born. They proved to be less individualistic. Rather, they reacted as a cohesive group, refusing adamantly, with loud calls of distress, to be separated from one another.

In another experiment that was filmed in 1950, Lorenz divided a clutch of goose eggs into two groups, A and B. Group A was raised in perfectly normal conditions, the eggs being incubated by the mother and the goslings hatching in her presence, while the group B eggs were artificially incubated and the goslings hatched in the presence of the experimenter. Not long after hatching, the group A goslings began to follow their mother about, while those of group B, who had never seen their mother, faithfully followed Lorenz as he moved back and forth in front of the incubator. Later, Lorenz marked the goslings in order to be able to distinguish the two groups, and then he put them all together under a box. When the box was removed, the little goslings of group A clustered round their mother, while those of group B hastened to their adoptive mother, Lorenz!

The significance of these experiments is unmistakable: *filial and family ties are established very early.* These young birds inherit innate dispositions to orient themselves toward and follow their mother, or her substitute. They have an innate knowledge of specific calls and understand the meaning of auditory signals—calls of alarm, warning, flocking together, etc.—which Lorenz reproduced or imitated. But the object of their orientation, the creature to which they address these innate behavior patterns, is determined by *early experience* following immediately upon hatching. Ever since Lorenz first made these experiments, zoologists and psychologists have ceaselessly been delving into the effects of early experience on later behavior, in animals and also man. Lorenz pursued this type of experimentation with different species of precocial birds, and he was able to determine with great precision certain basic characteristics of the phenomenon (1935–1937). He called the process *Prägung*, which he himself translated into English as *imprinting*. The term conveys the sense of a sudden and profound impression, a very apt description of the nature of this phenomenon. The criteria established by Lorenz have been tested, evaluated, and verified by a number of writers (see general reviews and recent discussions in Sluckin, 1964 and Salzen, 1967). I shall summarize them briefly.

A duckling, gosling, or chick—that is, any young precocial bird—shows an *orientation and following response vis-à-vis the first moving object to which it is exposed after birth*. It is obvious that in natural surroundings, the mother, which incubates and hatches the eggs, is this first object. In experimental conditions, the mother may be replaced by a model or some other maternal substitute. Orientation and following responses have been observed toward such heteroclite objects as a cushion or a cardboard box pulled along the ground, a bouncing football, a passing dog. . . . Shape matters little, what is most important is movement. The young birds prefer an anomalous but mobile object to a perfect, but stuffed and immobile, specimen of their own species. A certain relative size must be respected. Lorenz's goslings became disoriented when he stood up; they followed him much better when he went on all fours. The goslings always kept Lorenz in a certain perspective: they kept at a greater distance when he was standing; they came closer when he was on hands and knees or when he was entering the water; when he was swimming, with only his head showing, they climbed right up and congregated on his back, exactly as if he were their mother. Auditory stimuli are also important. Shortly before hatching, a mother hen and her chicks already communicate by little calls, before the chicks have even seen the light of day. Auditory stimuli, however, are not indispensable. They usually call attention to visual stimuli, and this combination advances the orienting and following response by a few hours. When a mallard's nest is on open ground, the ducklings can come and go near the nest, and they can get a good overall view of their mother. In this case, visual stimuli play a major role. When a mallard's nest is tucked into a tree trunk or perched upon a willow stump, the ducklings are more restricted; they remain under their mother and hence have a rather limited view of her. In this instance, auditory stimuli are very important. Furthermore, these ducklings which have had less visual stimulation are older when they follow behind their mother to make their first excursion beyond the nest.

The orientation response and the family bond which is sealed by it can only take place during a brief period soon after birth. Lorenz called this the *sensitive or critical period*. The starting time and duration of this period may vary according to the species. It occurs earlier in the moorhen than in the coot, which is a closely related species. This is only natural, however, since little moorhens tend to hop out of the nest much earlier than young coots (Hinde, Thorpe, and Vince, 1956). It is imperative that the family tie be secured before this great leap from the nest. In the mallard, the sensitive period occurs within the first thirty hours after hatching, more precisely, between the fifth and twenty-second hours, with maximum sensitivity between the thirteenth and sixteenth hours (Hess, 1958). *Within the limits of this sensitive period, a few minutes suffice to fix responses upon the first stimuli to appear.* The family bond is established at the same time. Although at first a young nestling is prepared to follow any stimulus whatever, it will, after a certain time, show reactions of fear, it will avoid and flee from new stimuli.

Researchers concur that the advent of fear is linked to the end of sensitivity. However, it is an open question whether an animal's sensitivity ceases at a certain age due to a spontaneous termination of the critical period; whether this sensitive period is blocked by the spontaneous appearance of fear reactions; or whether the appearance of fear merely follows upon the completion of fixation, indicating that responses are already fixed exclusively on a given stimulus.

Finally, *the effects of this phenomenon are lasting and irreversible.* For the rest of its life, an animal will retain the *imprint* of the animal or object to which it was first exposed. In all of its later social behavior, the animal will relate exclusively to animals of that first species, or to objects of that first class, without, however, changing the specific form of its behavior patterns.

Not until 1950 did this work of Lorenz begin to be known outside German-speaking countries. The immense interest it aroused blossomed into a multitude of studies and experiments. This research, for the most part analyses of basic principles, sought to check the validity of the criteria which Lorenz assigned to imprinting. Very often these critiques are not well founded, because the experiments on which they are based were carried out under conditions which diverged much too far from the phenomenon studied by Lorenz.

Some researchers used domestic ducks from commercial stock to study the orientation and following response. Lorenz had stressed that fixation took place very early and required only brief exposure, but nevertheless these researchers used ducklings whose age was known no more precisely than somewhere "between 32 and 55 hours old"—in any case too old to still be imprintable, the sensitive period having long since passed. Their conclusions regarding the critical period and the irreversibility of imprinting cannot be upheld, since the following reaction they observed, occurring at a later stage of development, is in no way the result of imprinting, but rather the result of normal learning.

Some investigators were surprised to observe that the following response ceased after a certain age, and they refused to accept irreversibility as a criterion of imprinting. As a matter of fact, the following response, an overt manifestation of the filial bond established by imprinting, is a response which is characteristic of a young animal, and the response has no more meaning, nor any reason to persist, after the animal has become independent from its mother. Experimenters using ducklings as subjects should realize that it is completely normal for the following response to disappear gradually as the ducklings reach the age of independence.

Still another researcher mixes up imprinting and the following response. He thinks he is working on the level of the phenomenon, when in fact he is dealing with only one of its consequences; he confuses the total fixation phenomenon with one of its manifestations, the following reaction.

On the other hand, the term imprinting has been much abused, particularly in the field of psychology. It should be pointed out that at present the existence of imprinting, as defined and characterized by Lorenz, has been experimentally proven in only a very few species, in precocial birds such as the Anatidae, Rallidae, and Gallinae, and also in certain mammals, the ungulates in particular. There is an imprinting like process which exists in altricial birds and in apes and monkeys, animals which are much more dependent on their parents during their early life, but it occurs at a later date and extends over a longer period (Thorpe, 1964; Sluckin, 1964; Salzen, 1967). In these cases, the term *impregnation*, or permeation, might be preferable.

Most species of fish have little or no role in caring for their progeny. There are, however, a few exceptions, cichlid fishes for example, where parents and offspring form a united family group. Cichlids can be divided into two main groups according to their type of family organization. On the one hand, there are the substrate breeders, characterized by the biparental family in which male and female form a stable family unit. Both fish share in the defense of the territory, and there the nest is dug and the eggs are laid. The two parents take turns fanning the eggs, and after hatching they alternate in keeping the fry grouped together until the yolk sac has been absorbed. After the young fish can swim, both parents continue to protect them and keep them within the territory for a period of up to several weeks. On the other hand, there are the mouth breeders, characterized by a maternal family in which the female usually assumes full charge of the progeny. Her encounter with the male is brief, and immediately after fertilization she takes the eggs in her mouth and leaves the spawning ground. Mouth incubation lasts until the fry have absorbed the yolk sac and are able to swim. When the female expels the fry, they cluster round her in a dense cloud. Family cohesion lasts about two weeks and during this period the little fish follow their mother's every move. On occasion, she will take the fry into her mouth again when danger threatens.

The question arises whether, in the establishment of family ties, a difference exists between biparental cichlids and maternal cichlids which is comparable to the difference between precocial and altricial birds. To find the answer, young fry, raised in isolation from the egg stage, were exposed to mobile models representing a variety of parental substitutes. It was established that mouth breeder fry possess an innate disposition to fix upon and follow any animated maternal substitute at first sight, within the limits of a certain sensitive period. On the other hand, it was found that the family cohesion of young substrate breeders results more from their response to specific congregating signals given by the parents, than from any innate disposition to follow the first moving object encountered. They can learn to follow a model, but only after repeated exposures of longer duration.

Parallels do exist, then, in the establishment of family ties, and they reflect certain similarities in the life patterns of quite diverse animals. On the one hand, there are parallels between substrate breeder cichlids and altricial birds, and on the other hand, parallels between mouth breeder cichlids, precocial birds and hoofed mammals. In the first group, the family bond is established gradually. In the second group, it is fixed instantly, on first contact, so that the offspring are immediately capable of following the mother in her peregrinations (Destexhe-Gomez and Ruwet, 1967).

The research discussed so far has dealt with orientation and following responses which appear in young animals at an early age and are hence most susceptible to experimentation. From the beginning, however, Lorenz had specified that all of an animal's social and sexual responses become fixed on the animal or object to which it has been imprinted. When the jackdaw which Lorenz had raised by hand became adult, it favored him with all its sexual attentions. It happens that when a jackdaw courts a female, it brings her worms which it stuffs in her bill. Lorenz, the chosen fiancée, had to keep his mouth clamped shut and pinch his nose, but still the jackdaw tried to stuff worms in his ears. Lorenz has cited other examples of birds which had fixed all their sexual behavior on aberrant objects or on species quite unrelated to their own. One lovebird which had been raised with a ping-pong ball chose this as a sex partner, caressing it as if it were the head of another lovebird. A peacock, which out of necessity during the war had been raised in the heated turtle house of the Vienna zoo, attempted when grown to mate with one of the giant tortoises!

The irreversible nature of a sexual imprint has been clearly established. Lorenz raised a group of geese, taking the role first of their mother, then of their friend. These geese (group A) were later put together with conspecifics which had not been imprinted to Lorenz (group B). After several years, when the group A geese were offered a choice between Lorenz and group B geese, they still preferred to address their social behavior to Lorenz. Schein (1958) confirmed this irreversibility with turkeys. Five years after first being exposed to man, and with no reinforcement in the meantime, they still preferred man to other turkeys.

Accidental cases, but no less convincing, have occurred in the ornithological collections at the biological research center of Zwin. In 1965, torrential rains forced a stork to abandon its nest, and the little new-born storks were dying of hunger and cold. The sole survivor was taken under the wing of the head keeper. He raised it successfully and was considered in turn a mother, a sister, and a mate. In later years, when the young stork had grown up, it reserved all its social behavior and greeting ceremonies exclusively for the keeper. Perched atop the keeper's house, the stork used to throw back its head and clack its bill each time the keeper appeared. It completely ignored other storks, apparently unaware they were members of its own species. A white-eyed pochard, a diving duck, came to the collection at a very tender age and was placed among a group of surface-feeding teal. Today, this pochard mingles in their community betrothal ceremonies. At the peak of the excitement, it exhibits its own pochard courtship pattern, while the teal, for their part, intensify their introductory shaking, seesaws, water spouts, and high-and-short antics. This hapless pochard shows off its inherited courting patterns in front of all the female teal, an audience without the slightest appreciation of their meaning. Similarly, a male eider duckling was put in a pond of female mallards. As an adult, it related all its sexual behavior to these mallards and remained faithful even after several eider females had been introduced into the pond. At a tender and impressionable age, both the pochard and the eider duck had received the indelible imprint of another species.

Schutz (1965) and then De Lannoy (1967) made extensive studies of sexual imprinting in ducks. If a member of species A is raised with young ducks of species B, as an adult it will try to mate with members of species B. Imprinting which determines the orientation and following response, and imprinting which determines the fixation of sexual behavior, are usually interrelated and both take place shortly after birth. The sensitive period for sexual fixation, however, lasts much longer than the sensitive period for the following response. If ducklings are separated from their parents and raised together in a group, each is imprinted to its group companions, whether they belong to the same species or not. If a duckling is raised alone by a parent or a substitute, it fixes its response on the parental figure. If the parent is a female, such ducklings will later form normal pairs; if the parent is a male, they will later form homosexual associations. Furthermore, Schutz found that only male ducks are imprinted sexually. Females recognize innately the distinctly colored plumage of their male conspecifics. This knowledge is not gained by experience, nor can it be changed by experience.

However, since female ducks of different species look very much alike, it has been postulated that it would not be easy for males to develop an innate recognition precise enough to distinguish the characteristics of their female conspecifics, and hence this knowledge must be acquired. Furthermore, in the teal of Chile, where the two sexes both have dull protective coloration and look almost identical, the female can be imprinted, either to a parent or to a companion with which it is raised, and this fixes its sexual preferences on the species of the imprinting object. This does not explain, however, why geese show no sexual imprinting in either sex, since they too are not dimorphic, both sexes having the same neutral plumage. Nor, on the other hand, does it explain why the male shelduck can be sexually imprinted on females of other species, when its female conspecific has very brilliant coloration.

Clearly, caution must be exercised before generalizing on the existence of this phenomenon. But in cases where imprinting does occur, it is superimposed on instinct, orienting an animal's inherited responses to the imprinting object. The fixed action patterns of filial, social, and sexual behavior are inherited, but the particular stimuli which release them are acquired and fixed during critical periods, when innate releasing mechanisms are sensitized to these stimuli.

On the theoretical plane, the imprinting phenomenon is an excellent example of the complementary interrelationship between the innate and the acquired. On the practical side, its implications cannot be ignored by those who in the name of conservation would attempt to replenish certain species by raising broods in artificial incubators, or by those who persist in caring for young wounded, abandoned, or imperiled animals, only to release them later. The examples of the stork and the eider duck show that such animals, once imprinted to another species, are irretrievably lost with respect to their own species. Ornithologists concerned with wildlife protection have conceived a program to reintroduce certain species of birds into those areas of Europe where they have been eliminated by predators. According to this plan young birds of species A would be removed from their native nests in the Balkans. They would then be placed in the care of species B in Spain by distributing them singly into nests which already have young of about the same age. This operation entails a great risk that the sexual behavior of the adopted birds of species A would become fixed on the host species B, which would be tantamount to psychological sterilization. In order to be sexually and socially normal, these birds should be raised by their own parents or with companions of their own species.

Nor can the important effects of imprinting on the later social and sexual behavior of animals be treated indifferently by those who are concerned with the development of human behavior—especially since this process occurs not only in fish and birds, but in mammals as well, for imprintinglike phenomena have also been detected in goats, sheep, dogs, and monkeys. Social bonds are established very early during a sensitive period. If individuals are isolated during this critical period, and consequently fail to achieve fixation at this age, anomalies appear in their later behavior.

CHAPTER 6

SOCIAL BEHAVIOR OF ANIMALS

When one speaks of the social behavior of animals, one automatically thinks of highly structured insect societies. These have always intrigued the imagination. Some people view their organization as a model of efficiency. But the study of animal social behavior is not limited to these rather rigid communities. It encompasses all manner of contact which an individual animal establishes with members of the same species. Many invertebrates do not actively seek out conspecifics. Rather, the various individuals of a species simply seek the same ecological conditions, and it follows naturally that they come together in the same places. These passive aggregations can prove to be advantageous. They increase the chances of sexual products being combined or exchanged; they ensure better utilization of food and other environmental resources; they provide better protection from danger.

The animal kingdom runs the entire gamut of group behavior, showing every possible variety from the simplest to the most complex. The complexity of the social structure of a given species is closely tied in with the nature of its sexual life and the relative importance of its reproductive tasks. Most invertebrates do not form pairs for reproduction; they simply discharge sexual products into the milieu. In many species of insects, and in fish, amphibians, and reptiles as well, paired activity is strictly limited to the time necessary for fertilization: nuptial flights of insects, where two partners are momentarily joined in the air before separating definitively; brief encounters between frogs in a pond, where fertilization is quickly accomplished, then the egg clusters are abandoned in the reeds and the union is dissolved. These meetings must occur at a precise moment and the partners cannot afford to miss each other. In reality, animals possess both long- and short-term rhythms: pituitary rhythms regulating cyclic development of the gonads, alternation between daily periods of activity and rest, etc. These metabolic cycles, veritable internal clocks, are synchronized with the external rhythms of the seasons and the days. Since the various individuals of a species each adjust their own rhythms to the same temporal factors in the environment, the chances are excellent that they will set out in quest of one another at the same time—the same day, the same hour, and sometimes very nearly the same minute—and hence will come together at just the right moment. In species where the young are cared for over an extended period, interaction between the sexual partners and between parents and progeny becomes complex. This evolution of social behavior is evident in various groups of invertebrates as well as vertebrates.

Communal insect societies

All groups have evolved toward greater independence from the environment, toward behavior which relieves them from certain chemo-physical exigencies and other constraining pressures of the milieu. Along with the higher mammals, the social insects—bees, ants, termites

—show the most conspicuous marks of this evolution and have carried it to the most advanced stage.

The vital and functional unit is the community: beehive, ant hill, termite hill. The members of this society cooperate for the common good. They divide the labor among distinct castes, and specialization can reach such a degree that it is expressed by polymorphism, as manifested in the readily distinguishable worker, soldier, and reproductive termites. The individual is but a part of the whole and is not viable alone. Separated from the community, it dies, just as tissue separated from an organ degenerates. On the other hand, the total integration and subjection of the individual to the society make it possible for these insects to develop veritable civilizations, with such practices as herding, farming, gathering, stockpiling, warfare with other colonies, enslavement of the vanquished, etc. Some writers, such as Chauvin, go so far as to postulate that insect societies are able to achieve these complex activities through a sort of interconnection between individual insect brains. Each insect is of course very small, its miniscule nervous system comprises but a few cells, and its individual possibilities are limited. However, a pooling of individual nerve potentials on the level of the society would, in effect, constitute a sort of superorganism, which would have its own logic, its own metabolism, its own regulatory systems. This type of society is kindling a great deal of research, the most active schools being those of Von Frisch, Grassé, and Chauvin.

Bees have societies of varying complexity. Solitary bees do not form real colonies. A fertilized female builds her own nest in a burrow in the ground or in a crack in a rock, and she raises her own larvae on a mixture of honey and pollen. At the end of the summer, the female dies, but the offspring become torpid and survive the winter. Bumblebees form annual colonies. The founding female raises the first larvae herself. Female workers come from this brood, and they bring nectar and pollen to the burrow, storing it in rude tanks fashioned from earth and wax. At the end of the summer, fertile females are born, each of which will found a new colony after spending the winter in solitary hibernation. The original queen, however, dies, and her colony breaks up.

Honey bees *(Apis mellifera)* establish lasting colonies. Each colony is descended from and organized around a single queen. During the nuptial flight, this queen has received the semen of five to ten males; she has stored away enough sperm to fertilize her eggs for several years. From February or March until October, she will lay from 1500 to 2000 eggs every twenty-four hours. The first fertilized eggs produce workers, females with atrophied ovaries, and the hive will include 40,000 to 50,000 of these. In May and June, a few hundred unfertilized eggs are laid in somewhat larger cells which the workers have constructed. While laying eggs, the queen travels across the comb, and when the tip of her abdomen comes in contact with an outsize cell, she blocks the release of stored sperm as the eggs descend. These eggs develop without fertilization and produce males. Meanwhile, a number of larvae receive special food—royal jelly—exuded from glands in the pharynx and mandibles of workers. The larvae fed on royal jelly turn into fertile females with normally developed ovaries. In the nuptial flight, each of these females is followed by a suite of male courtiers. After mating, the young females duel for possession of the hive where only one queen is permitted to reign. As for the old queen, she makes ready for swarming, and leads off half the population of the colony with her. The males, being of no further use, are killed and ejected from the hive. Finally, just before the onset of winter, egg-laying ceases and the colony slows down its rhythm. The bees then huddle together for warmth, and, as long as the food reserves last, they can maintain a temperature of about 60° F. at the heart of a cluster.

During the period of intense activity in spring and summer, the workers organize into teams which divide the labor. Some clean the combs, others feed the larvae, still others produce wax and construct new cells. Guards posted at the entrance to the hive identify any strangers and promptly evict them from the colony. Workers in charge of ventilation are also stationed at the entrance as needed. They fan air into the hive to maintain a constant temperature of about 95° F. Field workers gather pollen and nectar, while foragers seek new food sources. Each worker usually performs all of these tasks in the course of its short life, thirty to forty days. For the first few days, it is a cleaner. At about six or seven days, that is, when the worker's cephalic glands produce royal jelly, it becomes a nurse for the young larvae. Subsequently, it becomes wax-maker and fanner at about twelve days, guard at about eighteen days, and finally, at three weeks of age, forager and gatherer. At one time it was thought that, as a worker grew older, it progressed from one task to another in a precise order and at a fixed pace. It has been found, however, that the proportion of bees engaged in one task or another varies, and their activity is actually determined by the needs of the colony at the moment. According to the circumstances, a worker's progress to new assignments may be accelerated or delayed. An experimenter can separate all the younger workers of a colony from the older ones by removing the combs from the hive at mid-day, when the honey-gatherers are in the fields. These combs, full of young workers, are transferred to a new hive, and later all the field workers will return to the original hive. At the new hive, the colony is now made up entirely of young workers. Some of these will undergo an accelerated development and be precociously transformed into foragers and gatherers. In the original hive, only old workers have been left behind. Some of these will remain in the hive and revert to being cleaners, wax-makers, guards, fanners, and even nurses, which will entail regeneration of their atrophied cephalic glands.

The most amazing example of the high level of cooperation and organization of a bee colony is the language of bees *(Bienensprache)*. For ever so long, beekeepers and entomologists had been intrigued by a tail-wagging dance which certain workers perform on the vertical combs of the hive. They had also noted with interest that a visit by a forager bee to a food source is soon followed by the arrival of a growing number of honey-gatherers. Von Frisch connected these two phenomena and solved the riddle of the waggle dance. At first his work aroused only incredulity and ridicule; but very quickly, Von Frisch was able to convince the skeptics.

He placed a bowl containing a sugar solution in a meadow, and then he waited patiently until this new source of food was discovered by a forager. While the bee was drinking from the edge of the bowl, the experimenter used a fine brush to mark its back with a little spot of dye. In the same way, he marked other early foragers which discovered the bowl. Before long, an increasing number of unmarked gatherers began to arrive. Soon they were coming in swarms, and then, as the supply of sweetened water dwindled, the number of visitors dropped off rapidly. Meanwhile, Von Frisch discovered that the marked bees were performing the famous waggle dance back in the hive. This dance attracted other workers which gathered round the performer and imitated the dance. The workers crowding in around the dancers were also given identifying marks. These bees proved to be the ones that flocked to the bowl of sugared water after its discovery by the foragers. Somehow, information indicating the location of the food source was being transmitted.

Von Frisch unraveled both the form and content of this message. If the bowl is placed within 100 meters, the forager performs a relatively simple round dance, giving no indication of direction. This dance conveys the message that there is a food source within a short distance from the hive. If the bowl is placed beyond 100 meters, the forager executes a figure-8 dance on the comb, providing information both as to distance and direction (Figure 24). It makes a short run in a straight line (the cross axis of the figure-8), wagging the tip of its abdomen in a rotary motion. Next, the bee circles to the right, repeats the straight leg, then circles to the left, and so on. The faster the dance and the wagging, the closer the food. This ratio is very exact and gives precise indications of distances up to 1000 meters. The orientation of the straight axis on the vertical comb provides directional information. If the bee moves upwards along this line, this means that the food is in the direction of the sun. If the dancer moves downwards, the food lies in the opposite direction. The axis of the figure-8 may be more or less inclined. The angle between the axis and the vertical corresponds to the angle between the path to the food and the horizontal projection of the sun. A 60° angle to the left of the vertical indicates that the route lies 60° to the left of the sun's projection.

It has been postulated that this dance may have originated from breaking down a double taxis into its component parts. When a bee flees from danger, it flies off in the direction of the sun. It manifests a negative geotaxis and a positive phototaxis simultaneously. In the dance, the vertical direction corresponds to the direction of the sun, that is, the direction away from gravity corresponds to the direction toward light. According to this hypothesis, a horizontal in the field—the projection of the sun—has been translated into a vertical on the comb.

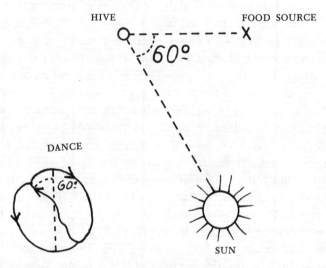

FIGURE 24. Waggle dance of the bee on the vertical comb of the hive to indicate the location of a food source by using the sun as a reference point. (After Von Frisch, *Aus dem Leben der Bienen*. Verstandliche Wissenschaft, neubearb. u. erg. 8 Aufl. Bd 1, Berlin—Heidelberg—New York, Springer, 1969.)

The sun, used as a reference point when indicating direction, moves across the sky during the course of the day. Bees fly slowly, and when food is abundant long dances are performed on the combs, sometimes for hours on end. In spite of the passage of time, the directional information remains extremely precise. For a distance of 800 meters, the margin of error is only a few degrees. The secret lies in the fact that the dancer in the darkness of the hive is constantly attuned to the changing position of the sun outside. In dances which continue for a long time, the bee gradually changes the inclination of the axis, so at any given moment, the directional indication for the food source is always correctly correlated to the position of the sun. Thanks to its metabolic clock, a bee has an acute sense of time. Its internal rhythm is synchronized with the twenty-four-hour day, and this enables the bee to keep track of the sun as it moves across the sky.

When preparations for swarming are underway, the foragers set out in quest of favorable locations for establishing the new colony. Upon returning to the hive, they provide information about the chosen site by this same waggle dance. Foragers and gatherers, then, are able to communicate by signals, conveying information about directions and distances to specific locations, places which the former have discovered and the latter will visit. In the entire animal kingdom, this is the only known example of communication which has the real rudiments of a clear and precise language.

The school of Von Frisch has specialized in the study of this language, while the French school inspired by Chauvin has been more concerned with the mechanisms which regulate the communal society of the hive. The scope of their research might almost be defined as "physiological determinism on the scale of the hive." They study the factors which regulate hive construction, food gathering, sexual ratios, division of labor. Particularly noteworthy is their finding that the central figure of the hive, the queen, exercises a dual power over the workers, a power which both attracts and inhibits. She attracts workers, and they congregate around her and care for her and the eggs. But the queen is also responsible for inhibiting ovarian development in these workers. The queen's dual power lies in her ability to produce certain chemical substances, and these have been isolated and identified. If the queen is removed from the hive, the workers will begin to develop ovaries. Those who trans-

form most quickly will in turn inhibit ovarian development in the others. Nevertheless, several queens are incompatible in one hive. They fight one another to the
death, and the sole survivor then commands the solicitous
attentions of the hive's entire population. Specialists in
the study of bees come to think of the colony as a viable
organic entity, dominating the individual members and
possessing its own social metabolism, its own regulatory
system.

Ants. The social behavior of ants is no less fascinating
to entomologists. All of the six thousand species are social. Rare fossil traces of their activity in ages past indicate that they have not changed very much in several
tens of millions of years. Contrary to bees, ant colonies
harbor numerous queens—up to two thousand—each one
capable of founding a new colony which could have a
population of three or four million.

Building upon a basic social framework, certain ants
have evolved quite specialized activities and social structures. The army ants of the tropics are nomadic. They
have no fixed ant hills, but establish temporary nests,
simple aboveground agglomerations of hordes of workers.
Tailor ants construct nests of leaves held together by silky
filaments. When a break occurs, chains of workers take
hold of the edges of the tear and pull it closed. Meanwhile, other teams have fetched some of their own larvae. Holding these adroitly in their forelegs, the workers
ply them back and forth like shuttles to mend the tear
with the sticky silk threads secreted by the larvae. Dairy
ants feed on the excreta of aphids. These ants raise
aphids in herds, putting them out to pasture on nutritive
plants and protecting them from enemies. On cool
nights, they bring the aphids into the ant hill. When
winter approaches, they collect aphid eggs and store

them away. Harvester ants store up grass seeds, and it is thought that some may actually cultivate grain. The leaf-cutter ants of America cultivate vast growths of fungus on leaf compost. This fungus is used to feed the larvae. Each species has its own specific fungus, and when a female sets out on her nuptial journey, she takes a strand of it in her mouth. When her first eggs are laid, a new fungus cultivation will be launched. Slave-maker ants are war-like and their ranks are made up mostly of soldiers. They pillage the nests of other species, eating the eggs and lar-vae. Some of the nymphs are carried off into captivity, and after hatching, these serve as slaves, performing the chores of the colony, including the care of the young.

Chauvin has made a recent discovery which indicates that different ant hills may establish communication with each other and maintain mutual relations by means of a network of paths which facilitate social intercourse. Rela-tionships of this nature—quite apart from war and en-slavement—are said to exist even between ants of differ-ent species.

Termites are far more ancient still than ants or bees. All of the species are social. Around the queen—an enor-mous laying machine—and her miniscule male, the col-ony is organized into castes of workers and soldiers. They feed on wood, the cellulose being predigested by micro-scopic intestinal protozoa. Termites also grow fungus on beds of wood particles, but unlike ants, they do not eat the fungus for itself, but rather for the lignite it assimi-lates. Termites are even more remarkable builders than ants or bees, and they erect edifices of enormous size and complexity. It passes all understanding how such tiny animals can coordinate their actions to achieve termite hills of several cubic meters, comprising pillars and arch-es, galleries, fungus beds, and a royal chamber.

Grassé has proposed the hypothesis that the edifice under construction is what stimulates the workers and directs their labors (the theory of stigmergy). When termites are experimentally put into a container supplied with building material, three successive phases of construction behavior can be observed. At first, the disoriented termites run about every which way in total anarchy: this is the phase of *noncoordination.* Next, some of them begin to shape little balls of earth and excrement which they arrange haphazardly and then abandon: this is the phase of *uncoordinated work,* in which one individual may destroy what its neighbor has just built. Eventually, several pellets happen to become stuck together by chance; these constitute a critical mass of stimuli which focuses the attention of all the workers; they add their own pellets to this mass, and very quickly the work takes shape and grows: this is the phase of *coordinated work.* Depending on how far the construction has progressed, the termites build walls, pillars, or arches. The product itself, then, by its size, shape, stage of completion, would seem to control its own construction.

Social Hierarchy and Territoriality

Social insects have carried their communal societies to extremes; indeed, the subjection of the individual to the community is absolute. Vertebrate societies are built on principles which are more within our ken, and their structure resembles our own social organization much more closely. Every vertebrate always maintains a security zone around itself, where no intrusions will be tolerated. This individual distance is the result of two conflicting tendencies, the tendency to attack and the tendency to avoid conspecifics. The ratio of attack to flight determines the spacing of individuals. This ratio varies from one species to another, and also within a species depending on the circumstances. Social spacing diminishes in the presence of a common danger. For example, a flight of starlings closes into a tight formation when a sparrow-hawk appears, but as soon as the predator leaves the scene, they resume their original spacing, each starling maintaining its distance. Individual distance also depends on where an encounter occurs. Aggression is greater where there is competition for sleeping sites, for choice perches, for access to drinking spots or food sources. Individual distance further depends on the season of year. If a sexual factor is added to a male's attack-flight conflict, its personal distance decreases vis-à-vis females and increases vis-à-vis other males. Last, it depends on individual familiarity. A stranger provokes more violent aggression than a neighbor or some other member of the group. This maintenance by individuals of a critical distance gives rise to two types of social organization: *social hierarchy* and *territoriality*.

Social hierarchy is the type of organization frequently found in nomadic animals, where relations between different group members are governed by the relative dominance or inferiority of the individuals. An animal which is more aggressive than the others maintains a zone of inviolability around itself wherever it goes. It preempts the best sites for eating and sleeping, and it tends to monopolize the females. It enforces respect for its right by force—using bill, teeth, or claws, as the case may be—and all the other individuals observe the established distance out of fear. This aggressive individual holds the "alpha" position, and just below it is a "beta" animal which also dominates the group, all save one. It submits to the despot, but distributes blows to all the others as underlings. Each and every individual has a rank order and at the very bottom of the hierarchy is the "omega" individual, dominated by all and with no one left below to dominate. This is a linear hierarchy. There are also examples of triangular hierarchy, in which subject A dominates B, which dominates C, which in turn dominates A. Each animal defends its rank by force—the pecking right maintains the pecking order (Schjelderup-Ebbe). Rank is relative and depends on size, age, sex, physical condition. Rank order can be changed by changing an animal's physiological condition. Also, an individual holding position x in one group, may occupy position $x+n$ or $x-n$ in another group.

Territoriality exists when an individual, a pair, or a group establish themselves in a particular place and deny other members of the same species access to it. Territoriality takes many different forms. Some territories are defended all year round, such as those of *Badis badis* fish, European robins, and hamsters. Most are temporary, however, and limited to the reproductive period. Territoriality has an advantage over social hierarchy in that each animal is master in its own domain. The animal becomes intimately acquainted with its environment, thoroughly exploring all the possibilities of its territory. It knows the area's resources and is familiar with its landmarks. But as soon as an animal leaves its own territory and penetrates that of its neighbor, it loses its dominance and falls to a position of inferiority.

It should be pointed out that hierarchy and territoriality can occur together in a wide variety of combinations. A territorial group can be organized hierarchically within their common territory. On the other hand, various members of a hierarchical group can acquire individual territories.

Whether an animal lives within a hierarchical group or maintains its own territory, the individual must assert its rights by aggressiveness. This aggressiveness is a sign of the vitality of the individual, but it carries risks for the species. It should not end in injury or death for one of the antagonists. It is advantageous to the species from the point of view of survival that aggressiveness not be translated into actual combat. Intraspecific combat, therefore, is essentially symbolic and incurs no bodily harm. The natural aggressiveness of conspecifics has been rendered innocuous by the development of gestures, structures, and auditory signals which express dominance or submission. Thanks to these displays and ornaments, the defense of rank or territory can be achieved by sham battles.

It is striking to note how often postures of threat and appeasement have forms that are the reverse of each other. A territorial male stickleback with a red belly threatens intruders by pointing its snout to the ground; a female or an intimidated male is neutral in color, and it assumes an appeasement posture in foreign territory by staying near the surface and pointing its snout up. A gull threatening an intruder shows its red bill and black mask full face; a gull appeasing a rival turns its head aside, hiding the red bill and black mask. A dominant dog or wolf bares its fangs; the vanquished exposes its throat to the victor. These signs of submission have the desirable effect of cooling aggression, whereas flight triggers hot pursuit.

These signals become very important when a sexual element enters into the balance between attack and flight. A male ready to mate has a tendency to chase off a female, while a frightened female is likely to flee from an aggressive male. The critical individual distance which determines attack or flight must be overcome in order for the partners to come together. They must learn to recognize each other and must activate mechanisms which consummate or redirect natural aggressiveness. This is the function of the manifold displays of courtship. These ritualizations bring about synchronization of the two partners and lead to their union.

Social and Family Organization of Fish

When one speaks of societies or families of fish, the layman is surprised because the picture that occurs to his mind's eye is schools of fish, such as herrings or sardines. The fact that certain species do care for their eggs and protect their young is, however, common knowledge among aquarium hobbyists. Indeed, fish show a varied spectrum of all kinds of social structure. The school is the simplest form. It is a large, uniform, cohesive collection, in which all the individuals are of the same size, the same age, the same color, and often the same sex. They are oriented in the same direction, they are spaced equidistantly, and they move as one in perfect synchronization. A school of fish is not a passive aggregation. Very early young fry show a drive to seek out conspecifics and to congregate. Equidistance between individuals is maintained by an equilibrium of motivation. The school is a type of societal organization often found in fish of the high seas, Clupeidae (herring), Mugilidae (mullet), Gadidae (cod). It is adapted to life in the open sea, facilitating movement from place to place, aiding in the search for food, and providing protection from predators. It is especially advantageous for reproduction, since it makes it possible for sexual partners to meet in the ocean's immensity. Actually, the coming together of the sexes is often more figurative than literal. Many fish living in schools have no courtship, no selection of a mate. In certain species, schools of males and females are superimposed when spawning. All the females expel their eggs simultaneously, and since the eggs are lighter than water, they rise toward the surface. As the eggs pass through the schools of males, they in turn expel their milt.

Certain species live in schools during periods of sexual quiescence and return to the coasts and shallows to stake out territories at the onset of the reproductive season. This practice is widespread among coastal fish. Many bottom-dwelling fish are territorial all year long and defend a hunting preserve as well as a reproductive area. The latter territory is defended by special behavior patterns, usually by the male, sometimes by the pair, and in rare cases by the female. Attacks are directed against members of the same species, especially against those in the same physiological conditon. The size of a territory is a species characteristic. Substrate breeders, such as *Tilapia* and sticklebacks, confine their defense chiefly to the layer of water near the bottom, while species which fashion a nest of air bubbles appropriate the layer of water near the surface.

Fish of the coral reefs are among the most aggressive of all territorial species. They live in quite shallow water, which is very pure and clear. They are patterned all over with gaudy designs of the most vivid colors. None of these harlequins will tolerate the proximity of another member of the same species, and an intruder is pursued ummercifully. Coloration is what triggers and orients the combat. Closely related but differently colored species are not attacked. Territoriality, then, is not related to competition for food, but is an integral part of reproductive behavior. Within a species, both sexes have the same markings, and a female that is not yet ready to mate is attacked with the same violence as a male. However, when a female that is ready to mate enters a territory, she is not attacked, because almost instantaneously her finery fades, her colors blanch, and in this way she appeases the proprietor's aggressiveness long enough for sexual union to take place.

After spawning, certain species pay no further attention to the eggs. Others protect the eggs, aerate them, chase off intruders, and keep the fry from scattering. In the paternal families of such fish as the stickleback and the *Tilapia macrocephala,* these duties are assumed by the male.

For certain species, the territory is merely a small area tantamount to a love nest where the male receives and courts female visitors. Males of the *Tilapia macrochir* make an encampment in shallow water and each one defends an area averaging about a meter in diameter. Whereas the males are restricted to their own quarters, females ready to spawn move about at will between different territories. They violate with impunity all the frontiers so fiercely defended against other males, and they are fervently courted by all the males staked out along their route (Ruwet, 1963).

When a female appears at a male's frontier, he approaches her, displaying his heightened ardor and combativeness. Inclined at 45°, with his snout pointed to the ground, he swims up to the female with rapid strokes of the tail fin, all the while exhibiting his long filamentlike spermatophores. He circles around the female, who in the meantime has crossed the border of the territory and has come to a halt at its center.

When a female appears in a densely populated spawning ground where territories are adjacent to one another, several males may approach her simultaneously. In this case, the female is the supreme arbiter of the competition. Its outcome is decided the moment she spontaneously crosses the boundary of a territory. The losing suitors show no further interest in her as long as she remains in the domain of a rival. She continues to act as a stimulus only for the owner of that particular territory.

The two fish thus arrive at the crucial stage in the formation of a pair—and how ephemeral this union will prove to be. At this point, the female may break off the engagement quite as spontaneously as she first appeared on the scene, and the male can do nothing to restrain her as she crosses over the border into the neighbor's domain where she will be greeted afresh with the same ceremony. On the other hand, the female may remain immobile, rooted to the spot. In this case, the male suddenly slides his snout under the female's belly and touches her genital papilla with his mouth. She then expels a little cluster of eggs which falls to the bottom. The female retires slightly to make room for the male, as he slides in over the eggs and turns on his side. The female advances again and touches her mouth to the extended filament of the male's genitals, and the male then expels his milt over the eggs. Now the male moves back, and the female comes forward once more to take the eggs one by one into her mouth where she will incubate them. It should be noted that in touching the male's genital filament, the female takes some sperm directly into her mouth before gathering the eggs. Actual fertilization, then, may even occur in the mother's mouth.

The series of events leading to fertilization is a chain reaction. If one step fails to occur, if one link is missing, the sequence comes to a halt. Therefore, it is desirable that the reaction be completed as smoothly and quickly as possible, and from the arrival of the female to her departure, not more than fifty or sixty seconds elapse.

The union of the pair now comes to an end, or it may temporarily be prolonged, or, more exactly, may be renewed, repeated. In fact, after the female has retrieved the fertilized eggs and stored them away in the recesses of her mouth, and the male has meantime resumed his ceremonial parade, there are two possibilities. The female may succumb anew to seduction by this same male, and the repetition of the chain reaction will result in the fertilization of a fresh batch of eggs. Or, the female may leave, perhaps spontaneously, perhaps due to the pair being disturbed as often happens in a densely populated spawning ground. She will then enter a neighboring territory to be greeted by a ceremony identical in every detail to the one already described. As for the first male, he will resume his patrol along the borders of his domain.

Each male receives a succession of females in his territory and fertilizes the eggs of a good proportion of them. Similarly, in the course of her journey through the spawning ground, a female stops off in several territories, yielding to the courtship of several males in succession, so the eggs she incubates in her mouth come from several lots fertilized by different males. In *Tilapia macrochir* and related types (the parental behavior of these fish was described in Chapter 5), pairing is extremely transitory and unstable. Successive polygamy and polyandry is the general rule. Other species of *Tilapia* form genuine unions, durable and stable. Both partners share in the defense of the territory, in the aeration of the eggs which are deposited in a nest, and in the care of the fry.

In either case, both males and females adopt certain postures and display special color patterns which are characteristic of the motivations of the moment. These expressive movements have a communicative function. An aggressive territorial male is brilliantly colored and stays near the bottom, while an intimidated male pales and rises to the surface. A female approaching a territory wears neutral dress to signal a symbolic inferiority and thus pacify the aggressiveness of the territory owner. When caring for their young, parents acquire a pattern of sharp contrasts and they execute brusque movements, signals which are useful in assembling the fry.

Detailed studies of expressive movements and the color patterns which emphasize them can be pursued only in the laboratory. Using large aquariums, the experimenter can diversify and repeat at will the conditions under which social and family partners meet. Postures and color patterns are then catalogued and analyzed. Observations are backed up by still photos and films ensuring authentic and accurate documentation (studies of Voss and Ruwet).

Territoriality in Birds

In 1920, a comprehensive study by Eliott Howard, an English amateur ornithologist, first attracted the attention of professional researchers to the phenomenon of territoriality in birds. A multitude of studies followed, and they have shown that almost all birds are, to some extent, territorial.

The acquisition and defense of a territory is part and parcel of the reproductive cycle. A bird becomes territorial through the intervention of internal factors, initiated by the pituitary gland, and through the action of external factors, emanating from social partners and the environment. External factors include higher temperatures, longer days, seasonal plant growth, and the general reawakening of nature typical of the reproductive period.

Sedentary species, especially small perching songbirds, follow more or less the same general pattern in establishing a territory. As sexual motivation becomes more explicit, the male tends to separate himself from the flock he used to roam with. At the same time, he begins to acquire certain secondary sex characteristics affecting his plumage, song, bill coloration, etc. He frequents a particular area more often, stays for longer periods, and defends it against other conspecifics. In effect, the male restricts himself topographically, and the individual distance within which no intrusion is tolerated is then considerably extended to become a "territorial distance." He is absolute master within his own territory, but he loses this superiority as soon as he ventures beyond its borders. In migratory species, the males usually take off first and reach the nesting grounds before the females. By the time the latter arrive, the males have already divided up the terrain into a mosaic of adjoining territories.

The manifestations which accompany bird territoriality are among the most conspicuous in the animal kingdom. Each male claims possession of an area by ritualized visual and auditory signals. He posts himself prominently on a high perch whence he pours out his song, or, failing a perch, he makes territorial flights. Bird song is a secondary sex characteristic signalling dominance over a territory. There are other auditory manifestations as well which have exactly the same function. Snipes dive from high in the air, spreading out their tail feathers to the maximum; the wing beats direct a flow of air alternately over and under these feathers, setting up a loud chattering vibration. Woodpeckers choose dead limbs to hammer away at with their bills. The drumming of woodpeckers, the chatter of snipes, the songs of passerines, all these sounds signal the presence of a male, inviting females to approach and warning rival males to keep their distance.

These auditory manifestations are most highly developed in birds which live in relative seclusion, notably the nightingale. Visual manifestations, the displays used in courtship, have the closely related function of signaling the ownership and defense of a territory. These displays are most highly developed in species living in the open, and they are complemented by conspicuous structures and color patterns.

Territorial modalities are so varied that it is not easy to classify them, and it is difficult to make an all-inclusive definition except in very general terms. For Howard (1920), any defended area is a territory. According to Tinbergen (1939), the term applies only to any area which is defended by combat of a sexual nature. Hinde (1956) distinguishes four general types of territory.

The first type of territory consists of an extensive reproductive area which may average several thousand square meters. During the reproductive period, all the activities of a pair take place there—courtship, mating, nesting, raising the young. This area provides most of the food for the parents and young. Finches and warblers mark out this kind of territory, and it is also typical of the European robin. As Lack (1946) has shown, this robin is unusual in that it defends its territory all year long.

The second type consists of a more moderate-sized territory. Activities linked to reproduction take place there, but it does not furnish the major part of the food supply for parents and young. Oystercatchers, avocets, lapwings, sandpipers carve up their common nesting grounds into individual territories which average a few hundred square meters. At certain times of day, neighbors will confront each other on the boundary line between their respective domains and indulge in ritualized border fights. At other times, they gather peaceably on communal feeding grounds.

The third type of territory shrinks to a square meter or so. Birds that nest in colonies, such as cormorants, gannets, gulls, terns, and sea mews, limit their territorial claims to the area immediately surrounding the nest. Pairs are adjacent, almost within touching distance.

These birds obtain all their food outside the colony.

The last type of territory, characteristic of only a few species, is not used for nesting at all, but serves only as a meeting-place for sexual partners. The males gather in a common arena where each one marks out his own little courting ground, and the females join them to choose their transitory partners. Immediately after mating, the females leave the area to nest in seclusion, while the males continue their display. Species which have evolved this kind of behavior—ruffs, black grouse, sage grouse, prairie chickens—show extreme sexual dimorphism. The males are larger than the females and exhibit conspicuous ornaments and vivid plumage. They take no further part in the reproductive process, all the subsequent tasks falling to the females. The plumage of the females is dull and monochrome, blending in with the vegetation.

It is interesting to note that the roles of the sexual partners are completely reversed in the painted snipes of Africa and in the Arctic phalaropes. In these species, it is the female who selects the territory and defends it. She is larger than the male, and it is she who sports the nuptial finery and performs the courtship display. The male, on the other hand, is relatively small and drab, and it is his task to incubate the eggs and care for the young.

Questions arise as to how adapted territoriality is, and what selection pressures have led to its development and cause it to persist. Hinde has enumerated the advantages inherent in possession of a territory. Of particular importance is the fact that a territorial bird, by confining itself to a limited area, gains a thorough knowledge of the terrain and can make optimal use of its resources. Also, territoriality helps to assure the smooth and uninterrupted functioning of the sexual process. It facilitates the coming together of the sexes, promotes mating, and strengthens the partnership. Furthermore, it reduces the despotism of certain males, giving each individual a chance. In certain species, however, such as the great tit, the birds pair off before a territory is established. The male focuses his song on a female first, and later localizes it in a particular area. In this case, territoriality serves primarily to establish ownership of a nesting site and to assure its defense, and territorial combat revolves around the nest. Humming birds defend clumps of flowers, and this is an example of territoriality protecting access to a food supply.

Another major advantage is that the spacing of pairs in distinct territories limits the population density, and consequently, it prevents excessive interbreeding and also reduces the risk of epidemics. It has been ascertained (Huxley, 1934) that a bird defends his territory most vigorously when he is at its center, and his zeal diminishes progressively as he gets further from its center. When new arrivals settle in an area and territorial dimensions are reduced, territorial aggressiveness increases proportionately. Combativeness will eventually reach a point where no further encroachments will be tolerated. Surplus pairs must then establish themselves in fringe areas (Ruwet, 1959). The size of a territory is flexible and can be compressed to a certain extent, but there is a limit to the number of territories which can be established within a given area. Hence, the territorial instinct seems to be a factor limiting population density.

The advantages of territoriality are manifold, and it is not possible to attribute to it one over-all function which holds true in every case. It is very likely that the selection pressures which produced territorial behavior in birds vary in importance from species to species.

Societal Organization of Apes and Monkeys

If ever the behavior of a zoological group has been

studied intensively, it is certainly that of apes and monkeys. For a long time, in fact, researchers believed that the study of simians held the key to problems of human psychology, and apes and monkeys were considered the ideal animal subjects for psychological research. For more than fifty years, macaques and chimpanzees, raised in zoos, menageries, or laboratories, were trained and subjected to all sorts of tests. They were taught to count, to distinguish geometric shapes, to draw, and to paint. Their intellectual aptitudes were discussed at length and a monkey psychology was formulated.

The net results of all this work are extremely deceptive. It has been taken for granted that the behavior and abilities of monkeys could be compared or contrasted with those of man. The tests devised in the laboratory did not deal with the true nature of the animal. These tests brought out characteristics which under the circumstances of the experiments seemed to be innate, but which are in fact quite alien to a monkey. Animals were used whose natural behavior was completely unknown, and they were subjected to testing and manipulation which was perforce totally unrelated to reality. Finally, very questionable conclusions—for the reasons just set forth—were rashly extrapolated from a few subjects which were more or less tame, and hence likely abnormal. These conclusions were then applied to primates as a whole, without the slightest regard for the phylogenic, morphological, ecological, or behavioral diversity of this group which numbers not less than 250 species divided among 80 genera.

Since monkeys do constitute first-class animal material for studying the causes and mechanisms of both normal and pathological human behavior, and also for understanding the processes of humanization, it is beyond belief that researchers were so long content to utilize animals adulterated by domestication or captivity. Many years passed before they tried to get to know monkeys as they really are in their natural ecological surroundings.

C. R. Carpenter, an American, was the first to do so in 1931–1933 when he went into the field in the Panama Canal Zone to study communities of howler monkeys in the wild. Carpenter was the pioneer of a new breed of researchers who have spread out over the world, to Japan, India, central Africa, and Latin America, and who have collected a far greater amount of vital data in the last decade than their predecessors did during the previous fifty years (cf. De Vore).

This type of research demands a great deal of enthusiasm, courage, and abnegation. In 1959–1960, George Schaller, of the University of Wisconsin, set up housekeeping with his wife in a little hut in the Congo's *Albert National Park* where gorillas roam the forested mountain slopes. He was the first person in the world to attempt to approach these reputedly dangerous animals unarmed. For fear that his gestures might be interpreted as a threat, he never trained binoculars or a camera upon them. At the cost of infinite patience, Schaller was able to approach closer and closer, and as his meetings with the gorillas became more frequent and lasted longer, he gradually came to understand their social structures and their means of communication.

At about the same time, an Englishwoman, Jane Goodall, undertook the study of chimpanzees in a Tanzanian game reserve. By dint of great patience, she succeeded, after some fourteen months, in approaching to within ten to twelve meters of these timid animals, without arousing their anxiety or disrupting their behavior. Eventually, her presence came to be so thoroughly accepted by these wild animals that she could mingle freely with the troop, and she even took up delousing the heads of several chimpanzees who returned the favor!

During 1959 and 1960, Phyllis Jay, of the University of California, lived in the field in India for two years in order to study the behavior of langurs. Each day, from dawn to dusk, she tried to come closer, bit by bit, to these very timid monkeys. At length, she came to be accepted, and upon her morning arrival the females and young monkeys flocked around her, tugging at her skirt and seeking to pull her into their games. One day, seated in the midst of a group, she turned back abruptly to get a notebook from a bag behind her. In doing so, she accidentally bumped a young female who immediately fled from the scene. At their first meeting after this fortuitous incident, the female turned and showed her hindquarters, as if she expected to be mounted. This is the posture by which a langur appeases a dominant individual. Thus, by bumping the female inadvertently, Mrs. Jay had acquired a position in the langur social hierarchy. This indicates how thoroughly integrated she became with the monkeys she had set out to study. Other noteworthy researchers are Suzanne Ripley who spent fifteen months in the jungles of Ceylon, and Irven De Vore who passes a good part of his life in central Africa in the company of baboons—and there are ever so many more.

These approaches to animal behavior are outstanding, and to my mind they are the pinnacle of achievement to date. Exactly like Lorenz, when he imprinted jackdaws or geese and took the place of their brother or mother, these researchers are actually living their own experiments. They have a role in these experiments. They are integral parts of the research. In field work, however, the method is complicated by the difficulty of approaching these wild animals in regions which are often almost inaccessible.

But the efforts of these researchers yield results, for the observations which they make in the field—by integrating themselves into a family or social group, by following from day to day the spontaneous activities of a troop in its natural surroundings—are much more significant than all the intelligence tests carried out in the laboratory.

Greater understanding comes not so much from finding out that a chimpanzee can learn to perform some manual operation in a laboratory in order to get a reward, but rather from discovering that a wild chimpanzee uses tools, that it pokes twigs into termite hills to capture insects, uses leaves to clean its coat, and throws sticks and stones at intruders. Real fascination lies not in displays of pictures painted by a tame monkey in a menagerie, but rather in trying to find out if his wild cousin living in the bush sometimes whiles away the time by drawing in the dust.

We are greatly indebted to the field work which has been carried out over the last ten years, for today we have a much better knowledge of the societal structures found among monkeys and apes. These animals can now be grouped into four categories according to their social organization.

The majority of the prosimians, nocturnal or crepuscular, are solitary and quite unsocial. Males are aggressive and form only temporary associations with females. They will no longer tolerate a young animal once it has reached maturity.

Some arboreal and diurnal monkeys form permanent and monogamous family groups, including the male, the female, their young of that year, and sometimes their still immature offspring of previous years. This type of long-lasting relationship is found in the indri and in the lar gibbon.

Most arboreal and diurnal monkeys, however, form

much larger multifamily groups. These are loosely organized and do not conform to any single pattern. There is no established hierarchy. The composition of the troops is flexible, and an individual may leave one troop and become readily integrated into another. This type of community is found in the howler and spider monkeys of Latin America, in the colobus monkeys and chimpanzees of Africa, and in the langurs of India.

Finally, the ground-dwelling monkeys of the savannahs—Asiatic macaques and African baboons—form tightly knit social groups, highly organized, stable, and rigid, based on complex relationships of dominance and subordination, and including several classes which have their own rank order. The entire group is under the leadership of a dominant male, and he is surrounded by an entourage of adult males of various ranks. Grouped around this central hierarchy are the adult females, which the leader tends to monopolize, and they have their own rank order. Included in this caste are the dependent young which accompany their mothers. Subadult males constitute the class on the lowest rung of the hierarchy. In some species, dominant males rank above all females, while in others, certain males may be outranked by more dominant females. Rank depends on age, size, strength, and aggressiveness. A dominant individual reserves for himself the choicest females, the best sleeping spots, the tastiest morsels of food. The others must wait their turn. The leader must constantly maintain and assert his superiority, or his position will be immediately contested.

In relations between dominant and subordinate individuals, between parents and offspring, simians possess no more elaborate means of communication than do other animals, the most gifted utilize some thirty different sounds, scarcely more than certain song birds, and the majority of apes and monkeys possess a repertoire about like that of the carnivores, rodents, and ungulates. In addition, simians, like other animals, use these sounds for the here-and-now, as interjections and exclamations. Their vocabulary is riveted to the present and the concrete, and it expresses the motivation of the moment.

Moods and motivation are also expressed by a repertoire of postures and gestures, especially facial expressions, but these are neither more numerous nor more accentuated than those of dogs and cats, for example. These gestures, calls, and facial expressions are all signals whose meaning is understood by conspecifics. They are often intensified or reinforced by conspicuous structures or coloration, such as the colorful calloused hindquarters and large decorative noses of baboons and mandrills. After long and patient observation of gorillas, Schaller came to understand their changing facial expressions right down to the slightest crinkling of the eyes. In studying their gestures, he had noted that a gorilla shakes its head vigorously as a sign of appeasement. One day, at a bend in a trail, he suddenly came eyeball to eyeball with an enormous male. Schaller shook his head with all the vigor at his command in order to convey appeasement, and the gorilla turned aside and moved off peacefully.

Field studies have brought back a great mass of data which I have only touched upon here. With this extensive knowledge of apes and monkeys, it is now possible to work out coherent programs of laboratory research based upon reality. The studies of Mason and Harlow (see Bourlière, 1965, and De Vore, 1965) on the role of parents and playmates in the socialization of primates are model examples of this type of research. It is known that a young monkey is dependent on its mother for a long time, and then it is placed in the company of other young monkeys of the same age. Under the wing of the family or community, these juveniles grow up together. In laboratory experiments, the behavior of rhesus monkeys born in captivity and raised in isolation from parents and playmates was compared with the normal behavior of the same species in the wild, and also with the behavior of a group of rhesus monkeys which had been captured and imported as subadults, that is, after the period of juvenile companionship. It was ascertained that the subjects raised in isolation have a full repertoire of inherited behavior patterns characteristic of the species. However, they make poor and inappropriate use of these behavior patterns when placed with other monkeys of the same age. On the sexual level, copulation is less prolonged, less successful, and less frequent, and it is performed in abnormal positions, on the part of both the male and the female. In their social behavior, monkeys raised in isolation prove to be aggressive; they participate less in mutual delousing, their dominant-subordinate relationships are not well established, and their position in the social hierarchy is unstable and poorly defined. They are less disciplined and sociable, and they do not integrate well. Some are truly deranged and they bite, claw, and injure one another. Females who manage to mate successfully later prove to be uninterested in their young, or, on the contrary, excessively domineering.

These various types of behavior are familiar—juvenile delinquency, social maladjustment, sexual aberration, insufficient or excessive maternal instinct. It would be unthinkable to carry out this kind of experiment on man, but the results obtained from simian subjects are extremely important and relevant to human psychology and the problems of learning. They show that social contacts during childhood and adolescence play a vital role in the assimilation of an individual into society.

One can understand the great interest now being shown in organizing interdisciplinary teams, uniting zoologists, psychologists, anthropologists, and psychiatrists. The task of the zoologist is to pursue and step up the study of primate behavior in the field. In addition, as members of these teams, zoologists must determine how compatible the research programs are with ecological and ethological realities, and it will be their responsibility to unravel and differentiate normal behavior patterns from those which are deviant.

Another consequence of close parental relationships and highly developed socialization of the young is that a habit acquired by one member of a group is propagated among the others. In order to study this phenomenon, Japanese researchers attracted monkeys into clearings by setting out food, some of which was foreign to the monkeys. The purpose of the experiment was to determine how rapidly the monkeys would become accustomed to eating the new food. Observations over a period of some ten years made it possible to follow the progress of new feeding habits through the community and to learn how these habits were transmitted. It seems that the young are more adaptable, but in a hierarchy, a new custom spreads very slowly from the bottom toward the top. On the other hand, the old are more conservative, but once a habit is acquired on an upper echelon of the hierarchy, it spreads down very rapidly toward the base, passing from the old and dominant to the young and subordinate, from parents to their offspring. In one of these communities, a young female, about a year and a half old, learned to wash sweet potatoes in the sea before eating them. She was imitated by her mother and her playmates and, eventually, the habit spread throughout the entire society where it has become a cultural characteristic. Only a few old males remain recalcitrant. This habit has now passed to a second generation. For the first time then, we find evidence that advanced socialization makes it possible to transmit acquired characteristics to succeeding generations.

CONCLUSION

Ethology is a young science. It began to develop but a few decades ago in reaction to an animal psychology which was divorced from reality and centered exclusively on the laboratory—the animal psychology of the mechanists, reflexologists, and behaviorists. Ethology marked a return to the naturalist traditions exemplified by Fabre. It has brought to light the extraordinary diversity of animal behavior, and consequently has exposed the hazards of hasty generalizations made in an attempt to formulate all-inclusive theories. By demonstrating the complementary relationships between instinct, taxis, and learning, ethology proved the vanity of the disputes which embroiled the various schools of animal psychology in the first half of the century.

From the beginning, a basic tenet of this new science has been a thorough biological knowledge of animal species, and its guiding principle has always been the study of animals in their ecological integrity. Ethology analyzes phenomena in their natural frame of reference, in their logical context and sequence. Any and every experimental intervention must be consistent with these realities. Ethological inquiry is developing and expanding in four major areas: it studies the causes and mechanisms of behavior;

it researches the function of behavior patterns and their survival value; it traces the evolution of behavior on the level of the species; it traces the development of behavior in the individual.

A few pre-eminent figures have determined the course of ethological development. Heinroth, Whitman, and Huxley were the pioneers. Lorenz guided ethology into the realm of true science. He, himself, best exemplifies the originality and effectiveness of his methods. He has achieved an unparalleled understanding of animals. He lives among them in intimate daily contact; he understands their signal codes, and he communicates to them and they to him; he penetrates their social structures, becomes himself an integral family member. Most recently, Tinbergen and Baerends have led ethology to new advances in field research; they have used the rigorous procedures of experimental inquiry in their field studies of behavioral factors and mechanisms. They have achieved a rare synthesis, combining the qualities and methods of the naturalist with those of the laboratory researcher.

At first, ethology was essentially descriptive, but very soon it expanded beyond such narrow confines. In studying the living animal in its logical environment, ethologists came to formulate certain concepts, to work out explanatory diagrams and models, and these in turn have given great impetus to physiological research on the level of behavior. Today, endocrinology and neurophysiology are valuable partners of ethology in the study of the innate mechanisms of behavior. In order to understand the relationship between an animal and its external environment, a knowledge of the animal's fields of perception is required. In this area of sensory capacities, the ethologist must be guided by the findings of the biophysicist, and, reciprocally, the latter's research should be guided by the former. Ethology and ecology each have their own objectives, but they also share areas of common interest. In order to study an animal in its logical surroundings, an ethologist must know how to respect the animal's ecological integrity, and how to fulfill the animal's ecological requirements when reconstituting its natural environment in the laboratory. On the other hand, animal behavior is one of the factors an ecologist must consider in studying population dynamics and animal-environment relationships. Very early, the comparative study of behavior contributed to taxonomy and phylogeny and utilized genetics, and today these overlapping fields of interest continue to be important facets of ethological research.

Ethology, then, has expanded, has reached out toward the other branches of biology, integrating their resources and coordinating their efforts in the study of behavior. Ethology has become a science of synthesis. It is no longer a merely descriptive study of animal habits. It has taken on such broad dimensions that today ethology may be defined as the *biology of behavior,* signifying its use of the methods and findings of all the disciplines of modern biology. Hence, further progress of the science of animal behavior requires the formation of interdisciplinary teams in which specialists in diverse fields—taxonomists, ecologists, geneticists, biophysicists, neurophysiologists, endocrinologists, etc.—join forces in a common effort with naturalist-ethologists. All must share an interest in animal behavior, and the coordinator of the team must make sure that the research program conforms to the four basic areas of ethological inquiry: cause, function, evolution, and development of behavior. The team organized by Baerends at the University of Groningen is a model to be emulated.

Ethology is an end in itself, but its findings should also be heeded by other disciplines. This holds true for all sciences and activities involving the use of living animals. It also holds true for all disciplines concerned with the study of human behavior, normal and pathological. For aside from all the acquisitions of our own species, aside from man's enhanced neural and psychic potential, we are, after all, fashioned from the same basic mold as the other vertebrates, and the fundamental laws which govern our behavior and theirs are identical. The study of behavioral mechanisms in fish and birds is of prime importance to us, and the experiments with monkeys on the role of socialization show how much can be gained through the combined efforts of ethologists, psychologists, psychiatrists, and sociologists.

I shall have achieved my goal if this book leads to a better understanding of ethology and its place in the concert of biological sciences, and if, consequently, students of human behavior come to realize that ethology offers more than mere analogies to their own disciplines, but indeed, a fertile field for common endeavor.

REFERENCES

Armstrong, E. A. (1947), *Bird Display and Behaviour*. Gloucester, Mass.: Peter Smith.
_____ (1963), *A Study of Bird Song*. New York: Oxford University Press.
Baerends, G. P. (1941), Fortpflanzungsverhalten und Orientierung der Grabwespe *Ammophila campestris*. *Tijdschr. Entomol.*, 84:68-275.
_____ (1950), Specialization in organs and movements with a releasing function. *Symp. IV Soc. Exp. Biol.*, 337-360.
_____ (1960), Het organisme in zijn gemeenschap: aard en herkomst der communicatiemiddelen bij dieren. *Werken Rectoraat R.U. Gent*, 3:64-84.
_____ (1962, La reconnaissance de l'œuf par le Goéland argenté. *Bull. Soc. Sc. de Bretagne*, 37(3-4):193-207.
_____ (1964), The research programme of the zoological laboratory of the University of Groningen. *Archives Néerland. de Zoologie*, 16(1):149-171.
_____ and Baerends-Van Roon, J.M. (1950), An introduction to the study of the ethology of Cichlid fishes. *Behaviour*, suppl no. 1, 1-242.
_____ and Blokzijl, G. J. (1963), Gedanken über das Entstehen von Formdivergenzen zwischen Homologen Signalhandlungen verwandter Arten. *Z. f. Tierpsychol.*, 20:517-528.
Bastock, M., Morris, D., and Moynihan, M. (1953), Some comments on conflict and thwarting in animals. *Behav.*, 6:66-84.
Beach, F. A. (1965), Sex and Behavior. New York.
Bierens de Haan, J. A. (1937), Uber den Begriff des Instinktes in der Tierpsychol. *Folia Biotheoretica* II—Instinctus, 1-16.

———— (1948), Animal psychology and the science of animal behaviour. *Behav.*, 1:71-80.

Blest, A. D. (1957), The function of eyespot patterns in the Leptdoptera. *Behaviour*, 11:209-256.

Bourlière, F. (1965), Réflexions sur la biologie sociale des primates. In: *La Biologie Acquistions récentes.* Centre International de Synthese, pp. 245-261.

Carmichael, L. (1926-27), The development of behaviour in vertebrates experimentally removed from the influence of external stimulation. *Psychol. Rev.*, 33:51-58; 34:34-47.

Chauvin, R. (1969), *Psychophysiologie*, Vol. 2. *Le Comportement Animal.* Paris: Masson.

Craig, W. (1918), Appetites and aversions as constituents of instincts. *Biol. Bull.*, 34:91-107.

Crane, J. (1949), Comparative biology of salticid spiders at Rancho Grande Venezuela: IV. An analysis of display. *Zoologica*, 34:159-214.

———— (1952), A comparative study of innate defensive behaviour in Trinitad Mantids (Orthoptera, Mantoidea). *Zoologica*, 37:259-294.

———— (1957), Basic patterns of display in fiddler crabs (Ocypodidae, Genus *Uca*). *Zoologica*, 42:69-82.

Cullen, E. (1957), Adaptations in the Kittiwake to cliff-nesting. *Ibis*, 99:275-302.

De Lannoy, J. (1967); Zur Prägung Instinkthandlungen. *Z. f. Tierpsychol.* 24:162-200.

Destexhe-Gomez, F. and Ruwet, J.-Cl. (1967), Impregnation et cohésion familiale chez les Tilapia (Poissons Cichlides). *Ann. Soc. Zool. Belg.*, 97:161-173.

De Vore, I., ed. (1965), Primate Behavior: Field Studies of Monkeys and Apes. New York: Holt, Rinehart & Winston.

———— and Eimerl, S. (1965), *Primates.* New York: Time-Life.

Dilger, W. C. (1962), The behavior of love-birds. *Scient. Amer.*, 206:88-98.

Eibl-Eibesfeldt, I. and Wickler, W. (1962), Ontogenese und Organisation von Verhaltensweisen. *Fortschritte der Zoologie*, 15:354-377.

Faber, A. (1953), Die Lant- und Gebärdensprache bei Insekten. Orthoptera I. *Mitt. Staatl. Mus. Naturk. Stuttgart*, 287:1-198.

Fabre, J. H. (1879-1908), *Souvenirs entomologiques*, 80th edition. Paris: Librarie Delagrave, 1923.

Frisch, K. von (1955), *Vie et Moeurs des Abeilles*. Paris: A. Michel.

Godefroid, J. (1968), Essai de mesure d'un comportement instinctif à l'aide du conditionnement operant. L'amassement chez le Hamster doré. Thèse non publiée. Liège.

Grassé, P. P. (1963), *Zoologie*, Vol. 1. *L'Ethologie ou Science du Comportement*. Paris: Encyclopédie la Pléiade.

Heinroth (1911), Beitrage zur Biologie namentlich Ethologie und Psychologie der Anatiden. *Verh. Inter. Ornith. Kongr.*, 589-702.

Hess, E. H. (1958), Imprinting in Animals. *Scient. Amer.*, 198:81-90.

Hess, W. R. (1943), In: *Helvetica Physiologica Acta*, Vols. 1 and 2.

Hinde, R. A. (1956), Territory review. *Ibis*, 98:340.

―――― (1966), *Animal Behavior: A Synthesis of Ethology and Comparative Psychology*. New York: McGraw-Hill.

―――― Thorpe, W. H., and Vince, M. A. (1956), The following response of young coots and moorhens. *Behaviour*, 11:214-242.

Holst, E, von and St-Paul, U. von (1962), Electrically controlled behaviour. *Scient. Amer.*, 206:50-59.

Howard, E. (1920), *Territory in Bird Life*. New York: Atheneum.

Huxley, J. S. (1914), The courtship behaviour of the great crested grebe *(Podiceps cristatus):* with an addition to the theory of sexual selection. *Proc. Zool. Soc. London*, 491-562.

―――― (1923), Courtship activities in the Red-throated diver *(Colymbus stellatus)*; together with a discussion on the evolution of courtship in birds. *Journ. Limn. Soc. London*, 35:253-291.

―――― (1934), A natural experiment on the territorial instinct. *British Birds*, 27:270-277.

―――― (1954), *Evolution as a Process*. London: Allen & Unwin.

―――― (1963), Lorenzian ethology. *Z. f. Tierpsychol.*, 20: 401-409.

―――― (1966), *A Discussion on Ritualization of Behaviour in Animals and Man*. Philosophical Transactions of the Royal Society of London. Series B. Biological Sciences, no. 772, vol. 251, pp. 247-526.

Iersel, J. J., Van (1953), An analysis of parental behaviour of the male three-spined stickleback. *Behav.*, supp. 3.

———— and Bol. A. C. (1958), Preening of two tern species. A study on displacement activities. *Behav.* 13:1-88.

Jacobs, W. (1953), Verhaltensbiologische studien and Feldheuschrecken. *Z. f. Tierpsychol.*, 1:1-228.

Janowitz and Grossman (1949), Some factors affecting food intake of normal dogs and dogs with œsophagotomy and gastric fistulas. *Amer. J. Physiol.*, 159:143-148.

Jennings, H. S. (1906), *Behavior of the Lower Vertebrates.* New York.

Jepsen, G. Mayr, E., Simpson, G. (1949), *Genetics, Paleontology and Evolution.* Princeton: Princeton University Press.

Johnsgard, P. A. (1965), *Handbook of Waterfowl Behavior.* Ithaca: Cornell University Press.

Koehler, O. (1963), Konrad Lorenz 60 Jahre. *Z. f. Tierpsychol.*, 20(4):385-401.

Kohler, W. (1927), *The Mentality of Apes.* New York: Random House.

Konishi, M. (1965), The role of auditory feed-back in the control of vocalization in the white-crowned sparrow. *Tierpsychol.*, 22:770-783.

Kortland, A. (1940), Wechselwirkung zwischen Instinkten. *Arch. Néerl. Zool.*, 4:443-520.

———— (1956), Aspects and prospects of the concept of instinct. Vicissitudes of the hierarchy theory. *Arch. Néerl. de Zoologie*, 11:155-284.

Lack, D. (1946), *The Life of the Robin.* London: Witherby.

Lehrman, D. (1953), A critique of Konrad Lorenz' theory of instinctive behaviour. *Quart. Rev. Biol.*, 28:337-363.

———— (1964), The reproductive behavior of ringdoves. *Scient. Amer.*, 188.

Loeb, J. (1918), *Forced Movements, Tropisms, and Animal Conduct.* Philadelphia: Lippincott.

Lorenz, K. (1935), Der Kumpan in der Umwelt des Vogels. *J. Ornith.*, 83:596-607.

———— (1937a), The companion in the birds' world. *Auk*, 54:245-273.

———— (1937b), Uber den Begriff der Instinkthandlungen. *Fol. Biothéor.*, 2:17-50.

———— (1941), Vergleichende Bewegungstudien an Anatiden. *Supp. J. Ornith.*, 89:194-294.

_____ (1950), Ethologie der Graugans *Anser anser. Encyclo-pedia cinematographica*, C 560.

_____ (1958), The evolution of behavior. *Scient. Amer.*, 199 (6):67–78.

_____ (1960), Prinzipien der vergleichenden Verhaltensforschung. *Fortschritte der Zoologie*, 12:265–294.

_____ (1961), Phylogenetische Anpassung und adaptive Modifikation der Verhaltens. *Z. f. Tierpsychol.*, 18:139–187.

_____ (1965), *Evolution and Modification of Behavior: A Critical Examination of the Concepts of "Learned" and "Innate" Elements of Behavior.* Chicago: University of Chicago Press.

_____ (1968), *Il Parlait avec les Mammifères, les Oiseaux et les Poissons.* Paris: Flammarion.

_____ (1969), *On Aggression.* New York: Harcourt, Brace, Jovanovich, 1966.

_____ and Tinbergen, N. (1938), Taxis und Instinkthandlung in der Eirollbewegung der Graugans. *Z. f. Tierpsychol.*, 2:1–29.

Manning, A. (1967), *An Introduction to Animal Behavior.* Reading, Mass.: Addison-Wesley.

Marler, P. R. and Hamilton, W. J. (1966), *Mechanisms of Animal Behavior.* New York: Wiley.

Marler, P. and Tamura, M. (1964): Culturally transmitted patterns of vocal behaviour in sparrows. *Sciences,* 146:1483–1486.

Marlier, G. (1959), Observations sur la biologie littorale du lac Tanganyika. *Rev. Zool. Bot. Afr.*, 59 (1–2):164–184.

Mayr, E. (1940), Speciation phenomena in birds. *Amer. Nat.*, 74:249–278.

Monfort-Braham, N. and Ruwet, J.-Cl. (1967), Les déclencheurs dans le comportement sexuel du *Pelmatochromis subocellatus Gunther* (Poisson Cichlide). *Ann. Soc. R. Zool. Belg.*, 97:131–159.

Morris, D. (1957), "Typical intensity" and its relation to the problem of ritualisation. *Behav.*, 11:1–12.

Moynihan, M. (1955), Some aspects of reproductive behaviour in the black-headed gull and related species. *Behav.*, suppl. 4.

Nice, M. M. (1937–1943), Studies on the life history of the song sparrow, 1 and 2. *Trans. Lin. Soc. New York,* vols. 4 & 6.

Pavlov, I. P. (1954), Selected Works. San Francisco: S. F. Book Imports.

Pelkwijk, J. J., ter and Tinbergen. N. (1937), Eine reitzbiologische Analyse einiger Verhaltensweisen von *Gasterosteus aculeatus* L. *Z. f. Tierpsychol.*, 1:193-201.

Richelle, M. (1966), *Le Conditionnement operant.* Neuchâtel: Delachaux & Niestlé.

Roeder, K. D. (1963), Ethology and neurophysiology. *Z. f. Tierpsychol.*, 20:434-445.

Rothenbuhler, W. C. (1964a), Behaviour genetics of nest cleaning in honey-bees. I. Responses of four inbred lines to disease-killed brood. *Animal Behav.*, 12:578-583.

———— (1964b), Behavior genetics of nest cleaning in honeybees. II. Responses of Fl and backcross generations to disease-killed brood. *Amer. Zoologist*, 4:111-123.

Ruwet, J.-Cl. (1959), Aspects du problème du cantonnement chez des oiseaux de la Réserve de Genk. *Le Gerfaut*, 49: 163-203.

———— (1963), Observations sur le comportement sexuel de *Tilapia macrochir* (Poissons Cichlides) au lac de retenue de la Lufira. *Behav.*, 20:242-250.

———— and Voss, J. (1966), L'étude des mouvements d'expression chez les *Tilapia* (Poissons Cichlides). *Bull. Soc. Sc. Liège*, 35:778-800.

Salzen, E. A. (1967), Imprinting in birds and primates. *Behav.*, 28:232-254.

Schein, M. W. and Hale, E. B. (1959), The effect of early social experience on male sexual behaviour of androgen injected turkeys. *Anim. Behav.*, 7:189-200.

Schleidt, W. (1962), Die historische Entwicklung der Begriffe "Angeborenes anslösendes Schema" und "Angeborener Auslösermechanismus" in der Ethologie. *Z. f. Tierpsychol.*, 21:235-256.

Schutz, F. (1965), Sexuelle Prägung bei Anatiden. *Z. f. Tierpsychol.*, 22:50-103.

Seitz, A. (1940-42); Die paarbildung bei einiger Cichliden. *Z. f. Tierpsychol.*, 4:40-84; 5:74-101.

Singer-Polignac (1956), Colloquium: *L'Instinct dans le Comportement des Animaux and de l'Homme.* Paris.

Skinner, B. F. (1938), *The Behavior of Organisms.* New York: Appleton-Century-Crofts.

_____ (1963), Behaviorism at fifty. *Science*, 140:951–958.

Sluckin, W. (1964), *Imprinting and Early Learning*. Chicago: Aldine.

Smith, S. (1947), *How to Study Birds*. London: Collins.

Tembrock, G. (1967), *Eléments de Psychologie animale*. Paris: Gauthier-Villars.

Thinès, G. (1966), *Psychologie des Animaux*. Brussels: Dessart.

Thorndike, E. L. (1898), *Animal Intelligence*. New York: Hafner.

Thorpe, W. H. (1960), Ethology as a new branch of biology. In: *Perspectives in Marine Biology*, ed. Buzzati-Traverso. Berkeley: University of California Press, pp. 411–428.

_____ (1961), *Bird Song*. New York: Cambridge University Press.

_____ (1964), *Learning and Instinct in Animals*. London: Methuen.

_____ (1965), Ethology and consciousness. In: *Semaine d'Etude sur Cerveau et Expérience consciente*. Pontifica Academia Scientiarum, pp. 705–709.

Tinbergen, N. (1939), Field observations of east-Greenland birds. II. The behaviour of snow bunting in spring. *Trans. Lin. Soc. New York*, vol. 5.

_____ (1948), Social releasers and the experimental method required for their study. *The Wilson Bull.*, 60(1):6–51.

_____ (1952a), Derived activities. Their causation, biological significance, origin and emancipation during evolution. *Quart. Rev. Biol.*, 27:1–32.

_____ (1952b), The curious behavior of the stickleback. *Scient. Amer.*, Dec. 414.

_____ (1953a), *Study of Instinct*. New York: Oxford University Press.

_____ (1953b), *The Herring Gull's World*. New York: Harper & Row.

_____ (1953c), *Social Behaviour in Animals*. London: Methuen.

_____ (1959), Comparative studies of the behaviour of gulls (Laridae): a progress report. *Behav.*, 15:1–70.

_____ (1960), The evolution of behavior in gulls. *Scient. Amer.*, 456.

Tinbergen, N. (1961), *Carnets d'un naturaliste*. Paris: Hachette.

_____ (1963), On aims and methods of ethology. *Z. f. Tierpsychol.*, 20(4):410–433.

———— (1966), *Animal Behavior.* New York: Time-Life.

———— and Kuenen, J. D. (1939), Uber die auslösenden und die richtunggebenden Reitzsituationen der Sperrbewegung von jungen Drosseln (*Turdus m. merula* L. und *T. e. ericetorum* Turton). *Z. f. Tierpsychol.*, 3:37–60.

———— Meeuse, B. J. A., Boerema, L. K. and Varossieau, W. W. (1942), Die Balz des Samtfalters, *Eumenis* (= Satyrus) *semele* (L). *Z. f. Tierpsychol.*, 5:182–226.

———— and Perdeck, A. C. (1950), On the stimulus situation releasing the begging response in the newly hatched herring gull chick (*Larus a. argentatus* Pontopp.). *Behaviour,* 3:1–38.

Tschantz, B. (1959), Zur Brutbiologie der Trottelume *(Uria aalge), Behav.,* 14:1–100.

Uexküll, J., von (1928), *Theoretische Biologie.* Berlin.

Verlaine, L. (1913-1939), *Oeuvres complètes.* Bibliothèque du Cercle des Entomologistes liégeois. Liège: Institut Van Beneden.

Viaud, G. (1951), *Les Tropismes.* Paris: Presses Universitaires de France.

Voss, J. and Ruwet, J.-Cl. (1966), Inventaire des mouvements d'expression chez les *Tilapia guineensis* et *macrochir* (Poissons Cichlides). *Ann. Soc. Zool. Belg.,* 96:145–188.

Wall, W., Von de (1963), Bewegungstudien an Anatinen. *J. Ornith.,* 104:1–15.

———— (1968), Le comportement des canards hybrides. *Ann. Soc. R. Zool. Belg.,* 98:125.

Watson, J. B. (1913), Psychology as the behaviorist views it. *Psychol. Rev.,* 20:158–177.

Weiss, P. (1941), Autonomous versus reflexogenus activity of the central nervous system. *Proc. Amer. Philos. Soc.,* 84:53–64.

Whitman, C. O. (1898), Animal Behavior. *Biol. Lect. Mar. Biol. Lab. Wood's Hole.* Boston. pp. 285–338.

———— (1919): The behavior of pigeons. *Carnegie Inst. Wash. Publ.,* 257(3):1–161.

Wickler, W. (1961), Okologie und Stammesgeschichte von Verhaltensweisen. *Fortschr. Zool.,* 13:303–365.

———— (1963), Zum Problem der Signalbildung, am Beispiel der Verhaltensmimikry zwischen *Aspidontus* und *Labroides.* *Z. f. Tierpsychol.,* 20:657–679.

———— (1968), *Le Mimétisme animal et végétal.* Paris: Hachette.